Lecture Notes in Computer Science   8440

Commenced Publication in 1973
Founding and Former Series Editors:
Gerhard Goos, Juris Hartmanis, and Jan van Leeuwen

Michael Chau   Hsinchun Chen
G. Alan Wang   Jau-Hwang Wang (Eds.)

Intelligence and Security Informatics

Pacific Asia Workshop, PAISI 2014
Tainan, Taiwan, May 13, 2014
Proceedings

Springer

Volume Editors

Michael Chau
University of Hong Kong, Hong Kong SAR
E-mail: mchau@business.hku.hk

Hsinchun Chen
The University of Arizona, Tucson, AZ, USA
E-mail: hchen@eller.arizona.edu

G. Alan Wang
Virginia Tech, Blacksburg, VA, USA
E-mail: alanwang@vt.edu

Jau-Hwang Wang
Central Police University, Takang Village, Taiwan, R.O.C.
E-mail: jwang@mail.cpu.edu.tw

ISSN 0302-9743   e-ISSN 1611-3349
ISBN 978-3-319-06676-9   e-ISBN 978-3-319-06677-6
DOI 10.1007/978-3-319-06677-6
Springer Cham Heidelberg New York Dordrecht London

Library of Congress Control Number: 2014936743

LNCS Sublibrary: SL 4 – Security and Cryptology

Typesetting: Camera-ready by author, data conversion by Scientific Publishing Services, Chennai, India

Printed on acid-free paper

Springer is part of Springer Science+Business Media (www.springer.com)

Preface

Intelligence and security informatics (ISI) is an interdisciplinary research area concerned with the study of the development and use of advanced information technologies and systems for national, international, and societal security-related applications. In the past few years, we have witnessed ISI experiencing tremendous growth and attracting significant interest involving academic researchers in related fields as well as practitioners from both government agencies and industry.

In 2006, the First Workshop on ISI was held in Singapore in conjunction with PAKDD, with most contributors and participants coming from the Pacific Asian region. The Second Pacific Asia Workshop on ISI, PAISI 2007, was held in Chengdu. Following that, the annual PAISI workshop was held in Taipei, Taiwan (2008), Bangkok, Thailand (2009), Hyderabad, India (2010), Beijing, China (2011, 2013), and Kuala Lumpur, Malaysia (2012).

Building on the momentum of these ISI meetings, we held PAISI 2014 together with PAKDD 2014 in Tainan, Taiwan, in May 2014. PAISI 2014 brought together researchers from a variety of fields and provided a stimulating forum for ISI researchers in Pacific Asia and other regions of the world to exchange ideas and report research progress. This volume of Springer's *Lecture Notes in Computer Science* contains research papers presented at PAISI 2014. It presents a significant view on regional data sets and case studies, including cybercrime, information security engineering, and text mining.

We wish to express our gratitude to all members of the Workshop Program Committee and additional reviewers who provided high-quality, constructive review comments within a tight schedule. Our special thanks go to the PAKDD 2014 Organizing Committee and workshop chairs. We would also like to acknowledge the excellent cooperation with Springer in the preparation of this volume. Last but not least, we thank all researchers in the ISI community for their strong and continuous support of the PAISI series and other related intelligence and security informatics research.

May 2014

Michael Chau
Hsinchun Chen
G. Alan Wang
Jau-Hwang Wang

Organization

Workshop Co-chairs

Michael Chau	The University of Hong Kong, Hong Kong
Hsinchun Chen	The University of Arizona, USA
G. Alan Wang	Virginia Tech, USA
Jau-Hwang Wang	Central Police University, Taiwan

Program Committee

Robert Weiping Chang	Central Police University, Taiwan
Kuo-Tay Chen	National Taiwan University, Taiwan
Uwe Glaesser	Simon Fraser University, Canada
Eul Gyu Im	Hanyang University, Korea
Da-Yu Kao	Central Police University, Taiwan
Siddharth Kaza	Towson University, USA
Paul W.H. Kwan	University of New England, Australia
Wai Lam	The Chinese University of Hong Kong, Hong Kong
Mark Last	Ben-Gurion University of the Negev, Israel
Ickjai Lee	James Cook University, Australia
You-Lu Liao	Central Police University, Taiwan
Xiaochen Li	The Chinese Academy of Sciences, China
Hongyan Liu	Tsinghua University, China
Hsin-Min Lu	National Taiwan University, Taiwan
Jun Luo	The Chinese Academy of Sciences, China
Xin Robert Luo	University of Minnesota, USA
Byron Marshall	Oregon State University, USA
Dorbin Ng	The Chinese University of Hong Kong, Hong Kong
Shaojie Qiao	Southwest Jiaotong University, China
Aixin Sun	Nanyang Technological University, Singapore
Paul Thompson	Dartmouth College, USA
Jennifer J. Xu	Bentley University, USA

Additional Reviewers

Nalbandyan, Narek
Yaghoubi Shahir, Hamed

Abstract of Invited Talk

Security Informatics Using Social Media Data

Ee-Peng Lim

School of Information Systems, Singapore Management University
`eplim@smu.edu.sg`

Abstract. Social media has become more diverse and pervasive in recent years due to unprecedented popular adoption of mobile and tablet devices. These new devices enable very fine grained tracking of users' attributes and behaviors as well as their relationships with other users. When we analyse the large volume of social media data, many interesting insights can be derived. Many of these insights help us to profile individual users and events in ways which were not possible before. In this talk, we describe a few interesting social media analytics works that address some important security informatics problems including societal-scale social media data sensing, user profiling, relationship mining and outlier detection. We will also highlight a few social media analytics tools that illustrate the security informatics capabilities.

Table of Contents

Rational Choice Observation of Malware Authors in Taiwan

Da-Yu Kao

Department of Information Management, Central Police University, Taoyuan, Taiwan 333
camel@mail.cpu.edu.tw

Abstract. Cybercrime is a significantly new phenomenon, which is facilitated by the internet. The internet not only makes it much easier for malware authors or hackers to bypass national boundaries, but also offer more sophisticated techniques to support malicious program creation for malware authors. This paper outlines an observation from Taiwan malware authors in past decades. The case description and author behavior in rational choice theories is discussed. The goal is to analyze their mentality and thinking patterns from a standpoint of rational choice theory. Profit emerges as a main factor in repeat offenders. They freely choose to write malicious programs after rational evaluation where the anticipated costs and the expected benefits are weighed against each other.

Keywords: Malware Author, Rational Choice Theory, Cybercrime, CIH Virus.

1 Introduction

Internet is a part of an information revolution that has offered opportunities for committing and being a victim of crime. Of approximately 50 thousands cases of cybercrime reported in Taiwan during the past decade, the majority were motivated by trivial disputes between online gamers. In Taiwan, 4 persons authored the malware (malicious software) earning the most publicity. They are known by their hacker pseudonyms: "CIH", "Birdspy", "Peep", and "keylog [1]." Although these noted authors are malware programmers, they have different life-course on crime. This paper aims to understand the extent between cyber technology and malware crime in Taiwan. The purpose of this study is to contribute to the literature by examining the links among rational choice and malware authors. More specifically, it aims to find out how advancements in cyber technology have impacted cybercrime and how often malware authors are relying on Information and Communication Technologies (ICTs) to commit the criminal offences.

In Section 2, the literature reviews of malware activities and Relational Choice Theory are discussed. Section 3 describes four Taiwan malware authors of CIH Virus, Birdspy Backdoor, Peep Trojan and Keylog Spyware. Discussions and analyses relational choice observation are presented in Section 4. The conclusion is drawn in Section 5.

M. Chau et al. (Eds.): PAISI 2014, LNCS 8440, pp. 1–7, 2014.
© Springer International Publishing Switzerland 2014

2   Literature Reviews

This section summarizes some of the literature from malware activities and rational choice theory, and explains its importance in the cybercrime analysis.

2.1   Malware Activities

Although the overwhelming majority of individuals on the internet are law-abiding citizens, the existence of cyber-attack presents us with an opportunity to enhance the foundation of our secure information infrastructure. As far as hackers are concerned, most of them are interested in hacking activities. However, malware authors are often interested in studying malicious codes and exploring system holes. They would like to share their achievements and obtain esteem from hackers who utilize their malware. When people express concerns about hacking activity, they have in mind a variety of different types of events: computer viruses, Trojans, worms, key loggers, Remote Administration Tools (RATs) and many other forms of malware attack. Malware is a general noun for malicious software that spreads between computers and their operations. Malware may be destructive deleting files or causing system crashes, but may also be used to steal personal data [9]. Some antivirus companies claim there are over 74,000 different malware in existence on malicious web domains [4, 8].

2.2   Relational Choice Theory

Rational choice theory, also known as rational action theory, is a framework for understanding and often formally modeling social and economic behavior [2, 9]. This theory is based on the "expected utility" principle from economic theory. Rational choice is an analysis of the thought or cognitive means from the environment. People make rational decisions based on choices they anticipate will maximize their profits or benefits and minimize the costs or losses [7]. Rational choice theory contains the following three components [3]. These components will be explored and discussed in Section 4.

(1) Beneficial Belief. The individual is likely to perform criminal offense that he or she believes are beneficial. This behavior implies that the individual makes a distinction between perceived costs and perceived benefits.

(2) Decision-Making Process. To explore the decision-making process, we must focus on specific crime, and capture the idiosyncrasies of different offense.

(3) Criminal Involvement. An important distinction is made between criminal involvement and criminal event. Criminal involvement is the process of acting upon information to become involved in a specific crime. Criminal event is what offenders use to participate in a specific crime.

3 Case Study

The internet not only makes it much easier for malware authors or hackers to bypass national boundaries, but also offers more sophisticated techniques to support malicious program creation for malware authors. Table 1 outlines an observation from Taiwan malware authors in past decades [1, 6, 8].

Table 1. Comparison between Taiwan Malware Authors

Item	Malware	1 CIH	2 Bisdspy	3 Peep	4 Keylog
Date	Design Date	1998.4-1998.5	1999.4-1999.8	2002.12	2004.4
	Arrest Date (News Date)	1999.4	2001.1	2004.5	2005.2
Individual Traits	Commit Age	23	23	30	17(1nd)/24(2nd)
	Motivation	Homework	Homework	Appy for M.S Entrance Exam	Profit (1st) / Challenge and Profit (2nd)
	Education Status (at Arrest)	University - Sophomore	University - Sophomore	B.S. Degree	Senior High School - eleven grade
	Repeated Offender	N/A	N/A	N/A	Yes
Malware Attribute	Malware Type	File Virus	Backdoor	Trojan	Spyware
	Malware Activity	Erase Data	Copy Files	Copy Files	Copy Accounts and Passwords
Profits / Benefits	Profits	N/A	NT $15,800	Tried but in vain	NT $ 6,000
	Estimated Costs	NT $ 250,000,000	NT $ 20,000	NT $ 50,000,000	NT $ 200,000

```
*=================================================================*
:*                    Modification History                        *
:*=================================================================*
*   v1.0      1. Create the Virus Program.                        *
*             2. The Virus Modifies IDT to Get Ring0 Privilege.   *
* 04/26/1998  3. Virus Code doesn't Reload into System.           *
*             4. Call IFSMgr_InstallFileSystemApiHook to Hook File System. *
*             5. Modifies Entry Point of IFSMgr_InstallFileSystemApiHook. *
*             6. When System Opens Existing PE File, the File will be *
*                Infected, and the File doesn't be Reinfected.    *
*             7. It is also Infected, even the File is Read-Only. *
*             8. When the File is Infected, the Modification Date and Time *
*                of the File also don't be Changed.               *
*             9. When My Virus Uses IFSMgr_Ring0_FileIO, it will not Call *
*                Previous FileSystemApiHook, it will Call the Function *
*                that the IFS Manager Would Normally Call to Implement *
*                this Particular I/O Request.                     *
*            10. The Virus Size is only 656 Bytes.                *
*=================================================================*
*   v1.1      1. Especially, the File that be Infected will not Increase *
*                it's Size...      ^_^                            *
* 05/15/1998  2. Hook and Modify Structured Exception Handing.    *
*                When Exception Error Occurs, Our OS System should be in *
*                Windows NT. So My Cute Virus will not Continue to Run, *
*                it will Jmup to Original Application to Run.      *
*             3. Use Better Algorithm, Reduce Virus Code Size.    *
*             4. The Virus "Basic" Size is only 796 Bytes.        *
*=================================================================*
*   v1.2      1. Kill All HardDisk, and BIOS... Super... Killer... *
*             2. Modify the Bug of v1.1                           *
* 05/21/1998  3. The Virus "Basic" Size is 1003 Bytes.            *
*=================================================================*
*   v1.3      1. Modify the Bug that WinZip Self-Extractor Occurs Error. *
*                So When Open WinZip Self-Extractor ==> Don't Infect it. *
* 05/24/1998  2. The Virus "Basic" Size is 1010 Bytes.            *
*=================================================================*
*   v1.4      1. Full Modify the Bug : WinZip Self-Extractor Occurs Error. *
*             2. Change the Date of Killing Computers.            *
* 05/31/1998  3. Modify Virus Version Copyright.                  *
*             4. The Virus "Basic" Size is 1019 Bytes.            *
*****************************************************************
```

Fig. 1. Modification History of CIH Virus

3.1   Sample Case of Malware Authors in Taiwan

CIH Virus in 1998-1999

The first example was 24-year-old Ing-Hau Chen who was a student at Tatung University, Taiwan in 1999. The name of the virus derived from his initials. The CIH virus, also known as Chernobyl or Spacefiller, was a Microsoft Windows 9x computer virus. According to Criminal Investigation Bureau, about 60 million computers were believed to be infected by the virus internationally, resulting in huge commercial damages on April 26, 1999. Pirated software, game software and music files were a prime source of infection [6]. Fig. 1 illustrates the modification history in CIH source codes [1].

Birdspy Backdoor in 1999-2001

The second example was 23-year-old Jeremy Chiu (aka Birdman) who was a student at National Central University, Taiwan in 2001. Bisdspy program was classified as a backdoor, which provided remote access to affected systems. BirdSpy usually acted by using several different ports for giving the hacker a remote access, connection and control to the victim computer. Birdspy backdoor properties were listed below:

• Log keystrokes;
• Connect itself to the internet;
• Send out application logs to the author;
• Hide activities from the victim computer;
• Stay resident in background environment;
• Allow remote user access, connection and control.

Peep Trojan in 2002-2004

The third example was 30-year-old Ping-an Wang who was an engineer from Kaohsiung, Taiwan in 2004. A Trojan horse called "Peep", and other malicious programs invaded hundreds of enterprise and government systems. That Peep was used by Chinese hackers to steal and destroy information on government-owned computers in Taiwan [1]. Peep can operate almost seamlessly across NATs and firewalls boundaries. It was believed that countless confidential electronic documents were stolen from compromised computers.

Keylog Spyware in 2004-2005

The fourth and last example was 17-year-old juvenile Huang (alias) who was a senior high school student from Taoyuan, Taiwan in 2005. He designed key-logging software in order to unlawfully gain access to the online game accounts and passwords. The investigation began with a complaint filed by one of the victims. In September 2011, law enforcement agents arrested him again while using Taipei MRT EasyCard system and making his sixth purchase at a convenience store. He was able to gain special access by wiring his home computer with a scanning machine to digitally alter the balance of three EasyCards on August 2011. Charged with breaching security systems, he face a maximum punishment of 10 years imprisonment and a fine of NT $200 million [1].

3.2 Similar Perspectives on Malware Authors

Case reviews of malware authors show differences in the particulars, but they also reveal common patterns. The cases of malware authors may vary as to interests, activities and organizations. The above four malware authors have different life-course on crime. We combine their life-backgrounds with extensive institutional records to ascertain the validity and reliability of the narratives. The life-backgrounds allow us to capture the diversity of life experiences and reveal the decision to commit a criminal offense [5]. Some similar perspectives on malware authors are listed below:

(1) Livelihood: Taiwan;
(2) Attributes: unmarried, young boy;
(3) Motivation: curiosity, interest, challenge, or money.

4 Discussions and Analyses

This section outlines theoretical research from the rational choice observation of Taiwan malware authors. The goal is to analyze their mentality and thinking patterns from the issues below [4, 5, 7, 9]: Individual Sense of Decision Making and Choice Component of Emotional Belief.

4.1 Individual Sense of Decision Making

Some malware authors have adopted this theory as their model of decision making. So do some malware authors. Even though one might consult others for their experiences with his decisions, each individual's circumstances are so highly particularized that the others' experiences may not be an appropriate guide to one's own best course of action. There is no widely accepted definition of rational choice theory, but there are two important senses in which the term is used [7, 9].

A Formal Sense: Transitive Preferences to Maximize the Profits or Benefits
The first sense in which the profession uses rational choice is more formal: People have transitive preferences and seek to maximize the utility that they derive from those preferences, subject to various constraints. The following two factors are included in their transitive preferences to maximize the profits or benefits for malware authors' choices [4, 9].
(1) Mediated Commencement. Their choices are mediated through a common medium, such as curiosity, interest, challenge, or money, which makes commensurability easier.
(2) Various Outcomes. Their choices of various outcomes are difficult to understand and have a variety of possible outcomes. The choices involve relatively straightforward comparisons in their mind.

An Informal Sense: Deliberative Alternatives to Minimize the Costs or Losses
The second is an informal sense: Choice is said to be rational when it is deliberative
and consistent. The decision maker has thought about what he or she will do and can
give a reasoned justification for the choice. The following two factors are included in
their deliberative alternatives to minimize the costs or losses for malware authors'
choices [5, 9].

(1) Frequent Choices. Their choices are frequent and routine. Serial crimes can
happen among malware authors. They have an opportunity to learn through repeated
actions.

(2) Transparent Problems. There is frequently a single or optimal decision. There
are systematic differences in transparent problems to cooperate by age or experience.
There may be some people who always obey the predictions of rational choice theory;
there may be some circumstances in which no one obeys those predictions. And there
may be more subtle differences.

4.2   Choice Component of Emotional Belief

Certain emotions enable malware authors to commit to actions that are in their long-
term interest instead of trying to maximize utility in each individual situation. After
calculating the efforts and costs of their efforts, malware authors believe they have
much to gain and little to lose, and they will not serve much prison time if they do get
caught.   Profit emerges as a main factor in repeat offenders. The rational choice on
malware authors is observed from the following three components [4, 7].

(1) Beneficial Belief. In their decision making, rationality of malware authors is
limited by the information they have, the cognitive limitations of their minds, and
the finite amount of time they have to make a decision. In Table 1, it is observed that
the authors of CIH, Birdspy, and Peep have designed their malware for academic
research. Most of them believed that that was beneficial to their studies.

(2) Decision-Making Process. From the observation of their criminal context, they
think to themselves that that is no big deal. All of them enjoy their sharing of malware
accomplishments. Some authors posted their programs on a Web, BBS, or FTP. Often
they would upload a document of related source codes or snapshots to prove their
claims. However, law enforcement agents considered their post behavior an enormous
security threat.

(3) Criminal Involvement. They would like to self-report their criminal involvement to law enforcement agents. Their self-report and official records reflect their
motive of creating, distributing and use of malware. A combination of both kinds of
records is an inclusive indicator of their criminal activities. When the fourth Keylog
author met the authorities, this boy openly told his story and did think that he was in
trouble. He did not learn lessons from that, and his behavior persisted for another six
years.

To sum up, these malware was not only designed for academic research, but Birs-
py, Peep and Keylog were also for profits. The authors of CIH, Birdspy and Keylog
are still active in Taiwan, and glad to share their new research to the public. Because
cybercrime science has been a significant agent of change for both the police and the
public, rational choice theory will be a necessary supplement to support the discovery

of evidence. It is believed to be a significant addition to the cyber criminology. Given that rational choice measures may not always account for all the variation of malware authors in individual behavior, additional research is necessary in the context of malware activities.

5   Conclusion

Cybercrime is a significantly new phenomenon, which is facilitated by the internet. The internet not only makes it much easier for malware authors or hackers to bypass national boundaries, but also offer more sophisticated techniques to support malicious program creation for malware authors. With the increased appearance of Advanced Persistent Threats (APTs), network intrusions have greatly threated the information society. Emotions provide a link between evolved preference and action. Computers had enriched the lives of malware authors, given their lives focus, and made their lives adventurous. They have enjoyed in turning dreams of endless possibilities into realities, and believed that everyone in society could benefit from experiencing such power. The follow-up research will pinpoint cybercrime threats, and analyze a network intrusion from a viewpoint of hackers' behavior analysis.

Acknowledgments. This research was partially supported by The Henry C. Lee Forensic Sciemnec Foundation.

References

1. Criminal Investigation Bureau, http://www.cib.gov.tw
2. Elster, J.: Explaining Social Behavior - More Nuts and Bolts for the Social Sciences, pp. 67–162. Cambridge University, Cambridge (2007)
3. Higgins, G.E.: Value and Choice: Examining Their Roles in Digital Piracy. In: Jaishankar, K. (ed.) Cyber Criminology Exploring Internet Crimes and Criminal Behavior, pp. 141–154. CRC Press (2011)
4. Hoffman, L.J.: Workshop Summary Report: Social Science, Computer Science, and Cyber security, Cyber Security Policy and Research Institute, p. 21. The George Washington University, Washington DC (2013)
5. Jacoby, J.E.: Classics of Criminology. Waveland, Illinois (2004)
6. Misra, G.: Permeance of ICT in Crime in India. Master Thesis, University of Twente, Enschede (2013)
7. Nosrati, M., Hariri, M., Shakarbeygi, A.: Computers and Internet: From a Criminological View. International Journal of Economy, Management and Social Sciences, 1–4 (2013)
8. Symantec Corporation, Internet Security Threat Report 2013. vol. 18, pp. 1–58 (2013)
9. Ulen, T.S.: Rational Choice Theory in Law and Economics, http://encyclo.findlaw.com/0710book.pdf

Knowledge Management in Cybercrime Investigation – A Case Study of Identifying Cybercrime Investigation Knowledge in Taiwan

Weiping Chang[1] and Peifang Chung[2]

[1] Department of Criminal Investigation, Central Police University, Taoyuan, Taiwan, 33304
una024@mail.cpu.edu.tw
[2] Information Management Office, Keelung City Police Bureau, Keelung, Taiwan, 20147
im701136@hotmail.com

Abstract. Law enforcement agencies are facing difficult challenges in fighting cybercrime in every country. The police need to learn vast skills and accumulate much knowledge to fight against cybercrime. Knowledge management (KM) is needed to fight against cybercrime. However, lack of knowledge management is one of the significant problems in law enforcement agencies in many countries. This paper is the continued study of fighting cybercrime - a KM perspective. Many law enforcement agencies don't know what knowledge they need to deal with cybercrime and what cybercrime investigation knowledge they have. This paper explores the knowledge needed and critical success factors for fighting cybercrime. In this paper, surveys are conducted in 20 police departments in Taiwan to identify the knowledge needed for fighting cybercrime. This study shows that an investigator with greater cybercrime investigation knowledge has better ability in cybercrime investigation and identities cybercrime investigation knowledge in Taiwan. Law enforcement agencies could identify and close the knowledge gap to successfully fight against cybercrime.

Keywords: knowledge management, cybercrime, crime investigation, cybercrime investigation, fighting cybercrime.

1   Introduction

Based on the statistics, there are a total of 4387 cybercrime cases in Taiwan in 2012. This is a remarkable decreased of 48.91% compared with the 8587 cases in 2011, that considerably increased by 123% compared to the 3851 cases in 2010 [15, 16]. As of 2012, general criminal detection rate is 84.03%, of which, violence crime clearance rate is 97.19% and 70.57% rate for theft cracked, that are all much higher than the rate of cybercrime clearance, 12.47%. The 2011 cybercrime detection rate is as low as 6.02% and in 2010 slightly higher to 17.42% [17], revealing the investigation difficulty of cybercrime case detection compared to other general criminal case. Moreover, to solve the cybercrime cases, the police need to learn more skills and knowledge, such as criminal investigation stages [2], Criminal Code, intrusion detection, computer

M. Chau et al. (Eds.): PAISI 2014, LNCS 8440, pp. 8–17, 2014.
© Springer International Publishing Switzerland 2014

forensics [21], etc. To manage the vast knowledge well, knowledge management is also needed. However, lack of knowledge management is one of the significant problems in law enforcement agencies in many countries. This paper is the continued study of fighting cybercrime - a KM perspective [3]. Most of law enforcement agencies often are not able to identify what knowledge they need to fight against cybercrime and what cybercrime investigation knowledge they have. This paper explores the knowledge needed and critical success factors for fighting cybercrime. To identify the knowledge needed for fighting cybercrime and what cybercrime investigation knowledge they have, surveys are conducted in 20 police departments in Taiwan. This study identities cybercrime investigation knowledge in Taiwan. After that, law enforcement agencies could capture, select, store, share, apply, and create the cybercrime investigation knowledge to manage the knowledge well [1].

2   Literature Review

2.1   Knowledge Management

The American Productivity and Quality Center (APQC) defines knowledge management as an emerging set of strategies and approaches to create, safeguard, and put to use a wide range of knowledge assets, such as people and information. Thus, these assets flow to the right people at the right time so that they can be applied to create more value to the organization [13]. Gupta et al. state that knowledge management is a process by which organizations are able to detect, select, organize, distribute and transmit vital information and experiences which would be used in activities like problem resolution, dynamic learning, strategic programming and decision making [9]. Kankanhalli and Kwok define knowledge management as a systemic and organizationally specified process for acquiring, organizing and communicating knowledge of employees so that other employees may make use of it to be more effective and productive in their work [12]. It will be seen from this that knowledge management can increases the ability to learn from its environment and to incorporate knowledge into the business process [14]. Therefore, knowledge management has been illustrated as a significant discipline in leading to positive performance in the organization. Without synchronization of knowledge management and core competencies, the organization would not succeed in long term survival and remain in competitive advantage [20].

There has been limited research and comment in cybercrime knowledge management. Gottschalk states that prevention of computer crime requires knowledge management [7]. Hinduja believes that leveraging knowledge from the past to address the future will improve computer crime investigations [10]. Cybercrime is requiring law enforcement departments in general and criminal investigators in particular to tailor a significant amount of their efforts toward successfully identifying, apprehending, and assisting in successful investigation and prosecution of perpetrators.

The identifying of knowledge and wisdom refers to the contentions of Housel and Bell [11], Stewart [21], Edvinsson [6], and the framework of knowledge management procedure is based on Bechman's theory: identifying, capturing, selecting, storing, sharing, applying, creating and selling [1]. Authors identify knowledge management procedure of policemen as follows:

(A) Knowledge Searching by Policemen
The police search for valuable and available knowledge inside and outside of the organization through information technology or other methods. For example, if the account used by the criminal is already known when investigating a cybercrime, a document can be sent to the ISP (Internet Service Provider) for account information, or to Telecomm for the name and address of the applicant.
(B) Knowledge Storing by Policemen
The police store knowledge in appropriate media by proper methods for the convenience of rapid searching, retrieving, and for easier updating to reconstruct contents.
(C) Knowledge Application by Policemen
The police must know how to exercise the authority given by law to enforce public affairs.
(D) Knowledge Sharing by Policemen
Knowledge sharing should consider "where to share", "who to share", "how to share" and "what to share". Knowledge sharing can even reach cross-unit members through information technology.

2.2   General Crime Investigation Knowledge

In 2000, Osterburg and Ward indicated that if an investigator has the capacity and relevant techniques (and knowledge) for investigating crimes, he/she will be more productive [18]. Criminal investigating capacity and techniques include crime scene processing and evidence collecting, investigation report writing, suspect arresting, assisting prosecutor in indictment, promoting witnesses to testify in courts, and so on. Gottschalk linked police knowledge to information systems [7]. Gultekin explored knowledge management in law enforcement, focusing on the POLNET system established by the Turkish National Police as a knowledge-sharing tool [8].

2.3   Cybercrime Investigation Knowledge

Casey stated that the procedure of cybercrime investigation is divided into 12 steps, such as accusation or accident warning, value evaluation, and standard procedure, affirmation or detainment of accident/crime scene are with or without search warrant preservation, recovery, result, simplification, system and search, analysis, report, persuasiveness and testimony, to roughly classify the knowledge required for investigating cybercrimes [2]. Many scholars have studied how to combat cybercrime. They developed many systems and tools [4, 5, 22, 23]. Chang proposed fighting cybercrime - a KM perspective and recommended the procedures of managing cybercrime investigation knowledge as follows [3]:

1) Determining what knowledge needs (KN) for the general crime investigation (GCI).
2) Determining what knowledge needs for the general cybercrime investigation (GCCI).
3) Determining what knowledge needs for each type of cybercrime investigation (CCI).
4) Assessing 1, 2, and 3 knowledge needs (KNs) and classifying them.
5) Identifying agency's cybercrime investigation knowledge (CCIK).

6) Finding the knowledge gap of cybercrime investigation between CCIK and KNs.
7) Filling the knowledge gap of cybercrime investigation.
The workflow chart of managing cybercrime investigation knowledge is as Fig. 1.

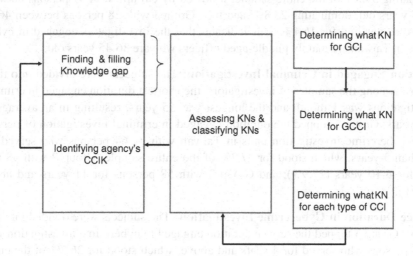

Fig. 1. The workflow chart of managing cybercrime investigation knowledge

3   Research Methodology

3.1   Questionnaire Survey

The questionnaire adopts a comprehensive survey on the units responsible for cyber-crime investigation in Criminal Investigation Bureau and in the city and county police departments in Taiwan. The survey is conducted by investigators in charge of cyber-crime cases. The structure of the questionnaire is mainly based on closed-end questions, with partially closed-end questions, and few open-end questions.

3.2   Study Subject

The main task force of fighting cybercrime is the 9th Investigation Brigade of Crimi-nal Investigation Bureau. There are few investigators in charge of cybercrime cases in most city and county police departments in Taiwan. However, some special task teams or units are established under criminal investigation corps in most police de-partments, such as Special Team for Computer Crimes Investigation under Criminal Investigation Corps of Taipei City Police Department.

3.3   Distribution and Collection of Questionnaire

214 questionnaires were distributed and 186 were collected, which was a response rate of 86.91%, and there were 172 effective questionnaires, of which the effective response rate was 80.37%.

3.4   Analysis on Surveyed Individuals' Background Information

Age: The subjects were divided into three groups with an average age of 39.42. Group 2 has the highest amount of respondents, 102 persons, between the ages of 36-45, dominating 59.3% of the entire sample; followed by Group1 with 41 persons between 26-35 years old, dominating 23.8%; and then Group3 with 28 persons between 46-55 years old, standing for 16.4%. This indicates that the investigators engaged in cyber-crimes in Taiwan are mainly middle-aged officers who are 36-45 years old.

Duration Engaged in Criminal Investigation: The subjects were divided into three groups. Among the samples of investigators, the shortest duration engaged in criminal investigations was 1 month and the longest was 35 years, resulting in an average of 8.75 years. Group 1 topped the duration engaged in criminal investigation of personnel in cybercrime investigation units in Taiwan with its 64 persons who served for less than 5 years, which stood for 37.2% of the entire sample; Group 2 with 48 persons for 6-10 years (27.9%); and Group 3 with 58 persons for 11 years and above (33.7%).

Service Duration in Cybercrime Investigation: The subjects were divided into five groups. Group 5 topped the service duration engaged in cybercrime investigation with its 46 persons who served for 4 years and above, which stood for 26.7% of the entire sample; followed by Group 1 with 45 persons for 1 year and under (26.2%); Group 2 with 39 persons for 1-2 years (22.7%); Group 3 with 20 persons for 2-3 years (11.6%); and Group 4 with 10 persons for 4 years and above (5.8%).

Highest Level of Education: The respondents were divided into four groups. Group 2 had the highest amount of people dominating in education level which was 71 persons with associate degrees, standing for 41.3% of the entire sample; followed by Group 3 with 57 persons with bachelor degrees (33.1%); Group 1 with 21 persons from high schools (12.2%); Group 4 with 20 persons with master degrees (11.6%). This indicates that over 70% of the personnel in cybercrime investigation units in Taiwan were graduates from Police College or Police University.

Cybercrime Cases Solved: The subjects were divided into five groups. 28 samples (16.3%) were missing values for this question, which was the second highest in individual background information. This indicates that this question is highly sensitive, and most test recipients refuse to answer it.

For this question, 27 persons were with the least value, i.e. 0 cases, and 1 person had the greatest value of 60 cases, resulting in an average of 6.69 cases. Among the samples of respondents, Group 1 had the most cybercrime cases solved, 91 persons solved 5 cases and under, standing for 52.9% of the entire sample; followed by Group 2 with 27 persons solving 6-10 cases (15.7%); Group 3 with 10 persons solving 11-15 cases (5.8%); Group 4 with 8 persons solving 16-20 cases (4.7%); and Group 5 with 8 persons solving 21 cases and above (4.7%). This indicates that over 50% of the personnel in cybercrime investigation units in Taiwan have solved 5 cases and under.

4 Data Analysis and Result

This study using Eoghan Casey's cybercrime investigative methodology [2] to design the questions on cybercrime investigation knowledge. Cybercrime investigation knowledge is divided into cybercrime investigation capacity, cybercrime problem solving ability, cybercrime learning application ability, cybercrime sharing ability, cybercrime technical ability. Cybercrime investigation capacity is divided into reading and interpretation ability, number of familiar command, number of familiar software tool, number of familiar field, number of certification and license, and number of familiar cybercrime type.

4.1 Cybercrime Investigation Knowledge

Service duration in cybercrime investigation is divided into Group a: 1 year and under, Group b: 1-2 years, Group c: 2-3 years, Group d: 3-4 years, and Group e: 4 years and above. Study results found that Cybercrime Investigation Capacity in Table 1 was of high significance ($P \leq 0.01$), indicating that the cybercrime investigation capacity of each group with different duration served in cybercrime investigation had significant differences. From these results we conclude that the cybercrime investigation capacities of investigators who have 4 years and above experience in cybercrime investigation are significantly better that those who have less than 4 years.

Item 5 "Cybercrime Technical Capability" in Table 1 was of high significance ($P \leq 0.001$), indicating that the cybercrime knowledge and technique of each group with different duration served in cybercrime investigation had significant differences. From these results we conclude that the cybercrime investigation technical abilities of investigators who have 4 years and above experience in cybercrime investigation are significantly better that those who have less than 4 years.

4.2 Technical Capability of Cybercrime Investigation

The "Reading and Interpretation Ability" between groups of service duration in cybercrime investigation was of high significance ($P \leq 0.001$), as shown in Table 2, indicating that the reading interpretation ability of each group with different duration served in cybercrime investigation had significant differences. From these results we conclude that the abilities of investigators, who have 4 years and above experience in cybercrime investigation, to read and interpret cybercrime related information are significantly better than that those who have less than 4 years.

"Number of Familiar Command" was of high significance ($P \leq 0.001$), indicating that the number of familiar commands between each group with different duration served in cybercrime investigation had significant differences. From these results we conclude that the number of cybercrime related commands known by investigators who have 4 years and above experience in cybercrime investigation is significantly higher than that those who have 1 year, 1-2 years, and 2-3 years in cybercrime investigation.

"Number of Familiar Fields" was of high significance ($P \leq 0.01$), indicating that the number of familiar fields between each group with different duration served in cybercrime investigation had significant differences. From these results we conclude that the number of cybercrime related fields that investigators, who have 4 years and above experience serving in cybercrime investigation, are significantly higher than that those with 1 year and less of experience.

"Number of Familiar Cybercrime Types" was of high significance ($P \leqq 0.001$), indicating that the number of familiar cybercrime types between each group with different duration served in cybercrime investigation had significant differences. From these results we conclude that the number of cybercrime case types that investigators who have more than 3 year experience in cybercrime investigation know how to process is significantly higher than that those who have 3 years and less experience in the investigation.

Table 1. Variance analysis table of service duration in cybercrime investigation in cybercrime investigation knowledge

Variable	Group	Effective Sample	Average	Standard Deviation	F Value	P Value	Post Hoc
Cybercrime Investigation Capacity	a	44	37.00	10.164	4.621	P≦0.01	e>a
	b	39	40.62	7.344			e>b
	c	19	39.00	8.756			
	d	10	42.80	7.131			e>c
	e	46	44.50	8.134			
Cybercrime Problem Solving Ability	a	44	10.95	2.496	2.023	P>0.05	
	b	37	10.54	1.952			
	c	20	11.25	2.731			
	d	10	10.90	2.685			
	e	46	11.96	2.366			
Cybercrime Learning Application Ability	a	45	11.29	2.464	1.731	P>0.05	
	b	39	11.40	1.636			
	c	20	11.20	3.088			
	d	10	11.50	1.958			
	e	46	12.35	1.969			
Cybercrime Sharing Ability	a	44	10.45	2.162	1.132	P>0.05	
	b	39	10.31	1.852			
	c	19	10.00	2.867			
	d	10	10.10	2.234			
	e	46	11.07	2.294			
Cybercrime Technical Ability	a	45	18.44	13.551	6.838	P≦0.001	e>a
	b	39	20.79	10.849			e>b
	c	20	19.80	14.685			
	d	10	28.50	18.039			e>c
	e	46	31.96	15.004			

Table 2. Variance analysis table of service duration in cybercrime investigation in Cybercrime Investigation Capacity

Variable	Group	Effective Sample	Average	Standard Deviation	F Value	P Value	Post Hoc
Reading & Interpretation Ability	a	44	1.70	1.887	5.829	P≤0.001	e>a
	b	36	1.94	1.756			
	c	20	1.90	2.222			e>b
	d	9	1.89	1.833			e>c
	e	46	3.34	1.882			
Number of Familiar Command	a	42	1.10	1.005	4.757	P≤0.001	e>a
	b	36	1.33	1.493			
	c	18	1.33	1.970			e>b
	d	9	1.56	1.878			e>c
	e	46	2.59	1.962			
Number of Familiar Software tool	a	43	1.51	2.324	1.851	P>0.05	
	b	36	1.81	2.724			
	c	18	1.67	2.401			
	d	10	2.60	4.088			
	e	41	3.02	3.174			
Number of Familiar Field	a	45	10.73	7.668	4.415	P≤0.01	d>a
	b	38	11.95	5.209			
	c	20	11.15	7.659			e>a
	d	10	15.90	9.012			e>b
	e	46	16.41	8.093			
Number of Certification and License	a	43	0.05	0.213	1.325	P>0.05	
	b	34	0.24	0.654			
	c	17	0.00	0.000			
	d	9	0.67	2.000			
	e	44	0.23	1.075			
Number of Familiar Cybercrime Type	a	44	3.61	2.982	8.369	P≤0.001	d>a
	b	38	4.37	2.353			
	c	20	4.05	2.762			e>a
	d	10	6.30	2.263			d>b
	e	46	6.61	2.879			

5   Conclusion and Future Work

Cybercrime is different from traditional crime. Cybercrime investigators need more knowledge to fight against cybercrime. As the police work environment becomes ever more complex, dynamic and stressful, police forces have to make efforts to improve the efficiency and effectiveness of police work [19]. Knowledge management is one of the ways to enhance the effectiveness of the police forces. Those who with more knowledge in cybercrime have higher ability and competition in fighting cybercrime, not only your co-workers but also cybercriminal. This study contributes to an understanding of the relationship between the cybercrime investigation knowledge and ability in cybercrime investigation. From this study we have learned that an investigator with greater cybercrime investigation knowledge has better ability in cybercrime investigation, and higher number of cybercrime cases solved. This study identities cybercrime investigation knowledge in Taiwan. Future works will be concentrated on filling the knowledge gap of cybercrime investigation.

Acknowledgement. The authors appreciate the anonymous referees for helpful comments and suggestions.

References

[1] Beckman, T.: Implementing the Knowledge Organization in Government. Paper and Presentation at the 10th National Conference on Federal Quality (1997)

[2] Casey, E.: Digital Evidence and Computer Crime, 2nd edn, pp. 92–113. Elsevier (2004)

[3] Chang, W.: Fighting Cybercrime: A KM Perspective. In: Chen, H., Chau, M., Li, S., Urs, S., Srinivasa, S., Wang, G.A. (eds.) PAISI 2010. LNCS, vol. 6122, pp. 28–30. Springer, Heidelberg (2010)

[4] Chen, H., Zeng, D., Atabakhsh, H., Wyzga, W., Schroeder, J.: COPLINK- Managing Law Enforcement Data and Knowledge. Communication of the ACM 46(1), 28–34 (2003)

[5] Donalds, C., OseiBryson, K.: Criminal Investigation Knowledge System: CRIKS. In: Proceedings of the 39th Hawaii International Conference on System Sciences, pp. 155–164 (2006)

[6] Edvinsson, L.: Developing intellectual capital at Skandia. Long Range Planning 30(3), 366 (1997)

[7] Gottschalk, P.: Information systems in police knowledge management. Electronic Government 4(2), 191–203 (2007)

[8] Gultekin, K.: Knowledge management and law enforcement: An examination of knowledge management strategies of the police information system (POLNET) in the Turkish National Police, Doctoral thesis of University of North Texas (2009)

[9] Gupta, B., lyer, L.S., Aronson, J.E.: Knowledge Management: practices and challenges. Industrial Management + Data Systems 100(1), 17–21 (2000)

[10] Hinduja, S.: Computer Crime Investigations in the United States: Leveraging Knowledge from the Past to Address the Future. International Journal of Cyber Criminology 1(1), 1–26 (2007)

[11] Housel, T., Bell, A.: Measuring and Managing Knowledge. McGraw Hill Companies (2001)

[12] Kankanhalli, T., Kwok, K.: Contributing Knowledge to Electronic Knowledge Repositories: An Empirical Investigation. MIS Quarterly 29(1), 113–145 (2005)

[13] Koenig, E.D., Srikantaiah, T.K.: Knowledge management lessons learned: what works and what doesn't. Medford, NJ: Information Today (2004)

[14] Liautaud, B., Hammond, M.: e-Business Intelligence Turning Information into Knowledge into Profit. McGraw-Hill, New York (2001)

[15] National Police Agency, 2012 Police Statistical News No. 5 (2013)

[16] National Police Agency, 2013 Police Statistical News No. 14 (2013)

[17] National Police Agency, 2013 Police Statistical News No. 3 (2013)

[18] Osterburg, J., Ward, R.: Criminal Investigation: A Method for Reconstructing the Past. Anderson Publishing Co. (2000)

[19] Seba, I., Rowley, J.: Knowledge management in UK police forces. Journal of Knowledge Management 14(4), 611–626 (2010)

[20] Shaghaei, N., Turgay, T.: Performance Improvement through Knowledge Management and Innovation in Educational Institutions: Teachers' Perception. GSTF Business Review 2(4), 143 (2013)

[21] Stewart, T.: Intellectual Capital: The New Wealth of Organization. Bantam Doubleday Dell Publishing, New York (1997)

[22] Zheng, R., Qin, Y., Huang, Z., Chen, H.: Authorship Analysis in Cybercrime Investigation. In: Chen, H., Miranda, R., Zeng, D.D., Demchak, C., Schroeder, J., Madhusudan, T. (eds.) ISI 2003. LNCS, vol. 2665, pp. 59–73. Springer, Heidelberg (2003)

[23] Zhou, Y., Qin, J., Lai, G., Reid, E., Chen, H.: Exploring the Dark Side of the Web: Collection and Analysis of U.S. Extremist Online Forums. In: Mehrotra, S., Zeng, D.D., Chen, H., Thuraisingham, B., Wang, F. (eds.) ISI 2006. LNCS, vol. 3975, pp. 621–626. Springer, Heidelberg (2006)

User Management in Information Security Engineering Environment ISEE

Yuichi Goto, Liqing Xu, Ning Zhang, and Jingde Cheng

Department of Information and Computer Science,
Saitama University, Saitama, 338-8570, Japan
{gotoh,xuliqing,zhang,cheng}@aise.ics.saitama-u.ac.jp

Abstract. An Information Security Engineering Environment (ISEE) based on ISO/IEC security standards has been proposed. ISEE integrates various tools such that its users can use these tools to ensure the whole security of their target information system at anytime consistently and continuously according to ISO/IEC security standards. In order to defend attacks and prevent damage beforehand, ISEE should provide users with some way to control user behavior by giving appropriate suggestions anticipatorily and actively. Such facility to control user behavior should be provided by a user management mechanism of ISEE, i.e., the user management mechanism of ISEE should deal with not only authentication, authorization, accounting or auditing, but also generating effective suggestions from records of user behavior anticipatorily, and informing the suggestions actively. Any traditional user management system is a storehouse of user data and works passively according to queries or transactions explicitly issued by its users and/or application programs, but has no active behavior to do something by itself. This paper presents an anticipatory user management mechanism of ISEE, as a new type of user management mechanism for SaaS-based cloud services with facility to control user behavior.

Keywords: Anticipatory user management, Information security engineering environment, Information security engineering cloud, ISO/IEC security standards.

1 Introduction

Information Security Engineering has many features that are intrinsically different from Software (Reliability) Engineering [9, 11]. The intrinsic difficulty to ensure the security of information systems is that crackers are active persons who can get knowledge and skills day after day and then continuously attack target information or software systems (information systems for short) always with new techniques. Thus, we have to improve our information systems and their managements to defend such new attacks continuously. Therefore, designers, developers, managers, administrators/end-users, and maintainers of information systems that require high security need continuous supports for their

M. Chau et al. (Eds.): PAISI 2014, LNCS 8440, pp. 18–34, 2014.
© Springer International Publishing Switzerland 2014

tasks to protect the system from crackers. The concept of an Information Security Engineering Environment (ISEE) was proposed [9, 11] because traditional software engineering environments are not adequate and effective for designing, developing, managing, and maintaining secure information systems. ISEE integrates various tools and provides comprehensive facilities for designers, developers, administrators/end-users, and maintainers of information systems such that they can use the tools and facilities to ensure the whole security of the target system anytime consistently and continuously according to ISO/IEC security standards.

As a platform to provide various services based on ISEE to various users, Information Security Engineering Cloud (ISEC) was proposed [54, 55]. To adjust the change in information technology and the evolution of security attacks, new security standards are published, already published security standards are revised, security best practices or anti-patterns [3] are found or revised. Therefore, it is hard for designers, developers, administrators/end-users, and maintainers of information systems to continuously follow all of security standards, security best practices, and anti-patterns related with own responsibility. Some of them may do not consider what is a better way to design, develop, manage, and maintain a target information system, what are defects in products of own tasks, and what action or activity leads to success or failure. In order to defend attacks and prevent damage beforehand, the ISEE should provide users with some way to control user behavior by giving appropriate suggestions anticipatorily and actively. For such active and anticipatory service, it demands to monitor whole user behavior on the ISEE, and preserve records of user behavior in the processable representation. It is easy to satisfy the requirements if the tools and facilities of ISEE are provided as a service, i.e., as a cloud service based on SaaS model [24].

ISEE and ISEC demand a common user management mechanism, and the user management mechanism should have not only facilities of authentication, authorization, accounting or auditing, but also facilities to manage records of user behavior, to generate suggestions to users from the records anticipatorily, and to inform the suggestions actively. Traditional user management mechanisms manage user identifiers and permission/roles, then control access to stored data and resources of a target information system, i.e., authentication and authorization. They also preserve records of user behavior, and monitor and analyze the records for accounting and auditing. Records of user behavior are undoubtedly pieces of information belonging to users. Hence, controlling user behavior is responsibility of user management as similar as controlling access to stored data and resources of in target information system. However, the traditional user management mechanism have not tried to control user behavior using records of user behavior [4, 14, 21, 22, 42–45, 59].

This paper presents an anticipatory user management mechanism for ISEE and ISEC, as a new type of user management mechanism for SaaS-based cloud services with facility to control user behavior. The anticipatory user management mechanism have not only facilities of authentication, authorization, accounting or auditing, but also facilities to manage records of user behavior, to generate suggestions

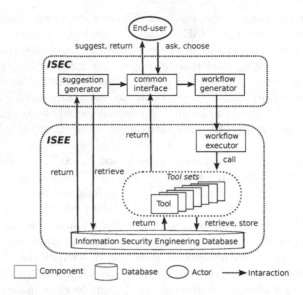

Fig. 1. Relationship between ISEE and ISEC

to users from the records anticipatorily, and to inform the suggestions actively. The rest of this paper is organized as follows. Section 2 gives brief explanation of previous and related works. Section 3 introduces anticipatory reasoning as a key technique to realize an anticipatory user management mechanism. Section 4 presents a requirement analysis of a user management mechanism of ISEE and ISEC, then presents an anticipatory user management mechanism and its implementation issues. Some concluding remarks are given in Section 5.

2   Previous and Related Works

2.1   Previous Works about ISEE and ISEC

Developing ISEE is an ongoing work [2, 25–28, 37, 49–52, 56–58]. The core component of ISEE is Information Security Engineering Database ISEDS [27, 37, 51]. ISEDS manages all of security ISO information security standards, such as ISO/IEC 15408 [29], ISO/IEC 18045 [30], ISO/IEC 27000 series [31–36], etc., and documents related with the standards, such as, security targets [29], protection profiles [29], etc., and provides integrated tools with the standards and documents. Each tool integrated into ISEE supports users to doing one or several tasks in software life cycle processes. Analysis of which tasks can be supported by software tools were done about tasks related with ISO/IEC 15408 [57, 58] and ISO/IEC 27000 series [50], and several tools were proposed and developed [2, 28, 52, 56]. However, there is no study about how to integrate and control the tools, although there are studies about appropriate sequence of execution of tasks from view point of information security engineering [26, 50] and relationship among tools [25, 58].

ISEC is also an going work [54, 55]. ISEC is SaaS-based application. The National Institute of Standards and Technology (NIST) defines cloud computing as "a model for enabling ubiquitous, convenient, on-demand network access to a shared pool of configurable computing resources (e.g., networks, servers, storage, applications, and services) that can be rapidly provisioned and released with minimal management effort or service provider interaction." [24] Cloud computing service model can be classified into three service models: software as a service (SaaS), platform as a service (PaaS), and infrastructure as a service (IaaS) [24]. A service model of ISEC is SaaS because one of roles of ISEC is a common user interface for end-users of ISEE. ISEC hides the complexity of usage of ISEE.

Figure 1 shows the relationship between ISEE and ISEC. ISEC takes what an end-user want to do from the end-user, then translates it into a workflow, i.e., how to do, according to published security standards and known best-practices. After that, ISEC sends the workflow to ISEE. ISEE chooses adequate tools to achieve the workflow, then calls the tools. Tools work automatically or interactively. On the other hand, ISEC gives suggestions according to known best practices and anti-patterns anticipatorily and actively.

2.2   Traditional User Management Mechanisms

There is no crisp definition of *user management* in papers about user management, but those researches deal with authentication, authorization, and auditing or accounting (well-known as AAA), the management of user profiles, and the management of user accounts. Role Based Access Control (RBAC) has been widely applied to authorize certain users to access certain data or resources [48]. Lin, et.al. [42] made RBAC more flexible to apply it into Web applications. Several researches [4, 14, 22, 45, 59] are about user management in grid computing environment. Those papers focus on integration between account management in virtual organizations and account management in the operating system in each node, which is a computer, to achieve unified AAA on a grid computing environment. Like the above user management in grid computing environment, Liu, et.al. [44] proposed a mechanism to unified authentication and authorization for a federated database system that is composed of a number of heterogeneous databases and is usually distributed. In cloud computing, Lin, et.al. [43] proposed an unified user management system that achieves that users can access different cloud services based on from SaaS to IaaS with the same identifier. On the other hand, Grzonkowski, et.al. [21] proposed a mechanism to share user profiles among web services. A user profile is personal data related with a specific user.

Through the above survey, we can summarize that traditional user management is to manage information for authentication, user profiles, information for access control to stored data and resources of the target system, and records of access to and use of the data and the resources. However, any traditional user management system is a storehouse of user data and works passively according to queries or transactions explicitly issued by its users and/or application programs, but has no active behavior to do something by itself.

2.3 Computing Anticipatory Systems

The main issue of this research is how to realize mechanism to generate effective suggestions from records of user behavior anticipatorily. There are studies about information systems that have the ability of anticipation. The concept of an anticipatory system first proposed by Rosen in 1980s [47]. Rosen considered that "an anticipatory system is one in which present change of state depends upon future circumstance, rather than merely on the present or past" and defined an anticipatory system as "a system containing a predictive model of itself and/or its environment, which allows it to change state at an instant in accord with the model's prediction to a latter instant." Dubois proposed the anticipatory system as a computing system, i.e., computing anticipatory system [15, 16].

On the other hand, from the viewpoints of software reliability engineering and information security engineering, what we need is really useful systems with anticipatorily predictive capability to take anticipation for forestalling disasters and attacks rather than the philosophical definition and intention of an anticipatory system. To develop anticipatory systems useful in the real world, Cheng proposed a new type of reactive systems, named "Anticipatory Reasoning-Reacting Systems," [5] as a certain class of computing anticipatory systems.

Anticipatory reasoning-reacting systems (ARRSs) were proposed as a new generation of reactive systems with high reliability and high security such that an ARRS predicts possible failures and attacks by detecting their omens and anticipatory reasoning about failures and attacks based on logic systems, empirical knowledge and detected omens, informs its users about possible failures and attacks, and performs some operations to defend the system from possible failures and attacks anticipatorily by itself. In other words, an ARRS is a reactive system with facility of prediction and decision-making.

3 Anticipation Mechanism Based on Forward Reasoning

3.1 Logic-Based Forward Reasoning

Anticipation is the action of taking into possession of some thing or things beforehand, or acting in advance so as preclude the action of another. It is a notion must relate to two parties such that the party taking anticipation acts in advance of a proper time earlier than the time when another party acts. To implement the facility of anticipation, we can naturally find following issues: how to predict future event or events, and how to take next actions. For the facilities, a prediction method and a decision-making method with forward reasoning based on strong relevant logic systems are proposed [6–8, 10, 38, 40].

Reasoning is the process of drawing new conclusions from given premises, which are already known facts or previously assumed hypotheses (Note that how to define the notion of 'new' formally and satisfactorily is still a difficult open problem until now). In general, a reasoning consists of a number of arguments (or inferences) in some order. An argument is a set of statements (or declarative sentences) of which one statement is intended as the conclusion, and one or more statements, called 'premises,' are intended to provide some evidence for the conclusion. An argument is a conclusion standing in relation to its supporting evidence. In an argument, a claim is being made that there is some sort of evidential relation between its premises and its conclusion: the conclusion is supposed to follow from the premises, or equivalently, the premises are supposed to entail the conclusion. Therefore, the correctness of an argument is a matter of the connection between its premises and its conclusion, and concerns the strength of the relation between them (Note that the correctness of an argument depends neither on whether the premises are really true or not, nor on whether the conclusion is really true or not). Thus, there are some fundamental questions: What is the criterion by which one can decide whether the conclusion of an argument or a reasoning really does follow from its premises or not? Is there the only one criterion, or are there many criteria? If there are many criteria, what are the intrinsic differences between them? It is logic that deals with the validity of argument and reasoning in general.

A logically valid reasoning is a reasoning such that its arguments are justified based on some logical validity criterion provided by a logic system in order to obtain correct conclusions (Note that here the term 'correct' does not necessarily mean 'true'). Today, there are so many different logic systems motivated by various philosophical considerations. As a result, a reasoning may be valid on one logical validity criterion but invalid on another.

In general, a formal logic system L consists of a formal language, called the object language and denoted by $F(L)$, which is the set of all well-formed formulas of L, and a logical consequence relation, denoted by meta-linguistic symbol $\vdash_L$, such that $P \subseteq F(L)$ and $c \in F(L)$, $P \vdash_L c$ means that within the frame work of L, c is valid conclusion of premises P, i.e., c validly follows from P. For a formal logic system $(F(L), \vdash_L)$, a logical theorem t is a formula of L such that $\phi \vdash_L t$ where ϕ is empty set. We use $Th(L)$ to denote the set of all logical theorems of L. $Th(L)$ is completely determined by the logical consequence relation $\vdash_L$. According to the representation of the logical consequence relation of a logic, the logic can be represented as a Hilbert style formal system, a Gentzen natural deduction system, a Gentzen sequent calculus system, or other type of formal system.

Let $(F(L), \vdash_L)$ be a formal logic system and $P \subseteq F(L)$ be a non-empty set of sentences (i.e. closed well-formed formulas). A formal theory with premises P based on L, called a L-theory with premises P and denoted by $T_L(P)$, is defined as $T_L(P) =_{df} Th(L) \cup Th_L^e(P)$, and $Th_L^e(P) =_{df} \{et | P \vdash_L et \text{ and } et \notin Th(L)\}$ where $Th(L)$ and $Th_L^e(P)$ are called the logical part and the empirical part of the formal theory, respectively, and any element of $Th_L^e(P)$ is called an empirical theorem of the formal theory. Figure 2 shows the relationship among $F(L)$, $Th(L)$, $Th_L^e(P)$, and $T_L(P)$.

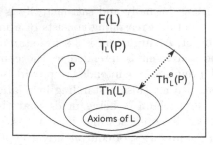

Fig. 2. *L*-theory with premises P

Automated reasoning is concerned with the execution of computer programs that assist in solving problems requiring reasoning. By adopting a suitable formal logic system for a target domain, we can do logically valid reasoning and get unknown or undecidable facts/hypotheses from empirical theorems that are well-known theories in a target domain. To do such logically valid reasoning automatically, a mechanism of automated reasoning is demanded. A forward reasoning engine is a computer program to automatically draw new conclusions by repeatedly applying inference rules to given premises and obtained conclusions until some previously specified conditions are satisfied. A facility to do reasoning automatically can be implemented by such forward reasoning engines and logic systems that are suitable for a target domain.

3.2   Anticipatory Reasoning and Reasoning about Actions

A method using anticipatory reasoning based on temporal relevant logics or 3D spatio-temporal relevant logics was proposed [6, 10]. *Prediction* is the action to make some future events known in advance, especially on the basis of special knowledge. It is a notion must relate to point of time to be considered as the reference time. For any prediction, both the predicted thing and its truth must be unknown before the completion of that prediction. An *anticipatory reasoning* is a reasoning to draw new, previously unknown and/or unrecognized conclusions about some future event or events whose occurrence and truth are uncertain at the point of time when the reasoning is being performed [6]. To represent, specify, verify and reason about various objects in the real world and relationships among them in the future, any ARRS needs a right fundamental logic system to provide a criterion of logical validity for anticipatory reasoning as well as formal representation and specification language. Temporal relevant logics and 3D spatio-temporal relevant logics are hopeful candidates of such right fundamental logic systems for ARRSs [6, 10]. Furthermore, to perform anticipatory reasoning automatically, an anticipatory reasoning engine was proposed and its prototype was implemented [8, 20, 46]. An anticipatory reasoning engine is a forward reasoning engine to perform anticipatory reasoning based on temporal relevant logics or 3D spatio-temporal relevant logics.

On the other hand, a decision-making method with reasoning about actions was proposed [38–40]. *An action* in a computing anticipatory system is a deed performed by the system such that, as a result of its functioning, a certain change of state occurs in the system. To take next actions, at first, a computing anticipatory system enumerates all actions that the system can perform in a predicted future situation as candidates of next actions, and then, the system chooses appropriate actions as next actions to defend the system from possible failures and attacks. The decision-making method uses reasoning about actions to enumerate candidates of next actions. *Reasoning about actions* in a computing anticipatory system is the process to draw new conclusions about actions in the system from some given premises, which are already known facts or previously assumed hypotheses concerning states of the system and its external environment [40]. Deontic relevant logics and temporal deontic relevant logics are adopted as hopeful candidates of right fundamental logic systems for reasoning about actions [7, 38, 40]. Furthermore, to perform reasoning about actions automatically, an action reasoning engine was proposed and its prototype was implemented [38, 40]. Like the anticipatory reasoning engine, an action reasoning engine is a forward reasoning engine to perform reasoning about actions based on deontic relevant logics or temporal deontic relevant logics.

The above anticipation mechanism based on forward reasoning, i.e., anticipatory reasoning and reasoning about action, is a hopeful mechanism to realize facility of generating effective suggestions from records of user behavior anticipatorily. In ARRSs, purpose of anticipatory reasoning is to deduce future event or events related with an ARRS, and purpose of reasoning about actions is to deduce candidate of next actions of an ARRS. However, we can change targets of both reasoning by using appropriate premises.

4 Anticipatory User Management Mechanism

4.1 Requirement Analysis

The requirements of the user management mechanism of ISEE and ISEC are as follows.

R1: *The user management mechanism should have a way to deal with authentication, authorization, and accounting or auditing (AAA), and management user profiles.* As mentioned above section 2.2, AAA and management of user profiles are standard facilities of user management mechanisms. Those facilities are necessary for both ISEE and ISEC.

R2: *The user management mechanism should have a way to adjust the change of security standards, security best practices, and security anti-patterns.* To adjust the change in information technology and the evolution of security attacks, security standards, security best practices, and security anti-patterns are created, revised, and expired. The change of standards causes change of a range of information assets, roles, responsibility of each role, best practices, and anti-patterns. Change of best practices or anti-patterns influences what the facility of ISEC suggests to users. One of the purposes of ISEE and ISEC is to fulfill the gap knowledge of designers, developers, administrators/end-users, and maintainers of information systems, and cutting-edge knowledge of information security engineering. Thus, the user management mechanism should have a way to adjust the change of security standards, security best practices, and security anti-patterns.

R3: *The user management mechanism should have a way to preserve user identifiers, user profiles, permission/roles, and records of user behavior securely.* Of course, user identifiers, user profiles, and permission/roles should be preserved securely because they are important information. Records of user behavior are not important and valuable when the number of the records is not so many. However, as a number of records of user behavior increases, the records becomes more and more important and valuable information. Thus, records of user behavior should be also preserved securely as well as user identifiers, user profiles, and permission/roles.

R4: *The user management mechanism should have a suggestion generating mechanism that can anticipatorily generate suggestions and actively give a target user them.* As mentioned in section 1, to defend attacks and prevent damage beforehand, it is necessary not only to manage records of user behavior, but also to control user behavior by giving appropriate suggestions anticipatorily and actively.

R5: *The suggestion generating mechanism should be able to generate suggestions for various targets in various level.* End-users of ISEE and ISEC are designers, developers, administrators/end-users, maintainers of information systems. On the other hand, there are best practices and anti-patterns in individual-level, team-level, organization-level, inter-organization-level. Thus, the suggestion generating mechanism should be also able to generate suggestions for design, development, management, and maintenance in individual-level, team-level, organization-level, inter-organization-level.

Traditional user management mechanisms have focused on only the requirement R1. There is no user management mechanism that satisfies R2 to R5.

4.2   Core of Anticipatory User Management Mechanism

Anticipatory user management is to manage user identifiers, user profiles, permission/roles, and records of user behavior, and to control not only access to stored data and resources of a target information system, but also user behavior by giving appropriate suggestions anticipatorily and actively. An anticipatory user management system (AUMS) is an information system that provides ISEE and ISEC with anticipatory user management. Figure 3 shows the relationship among AUMS, ISEE, and ISEC. AUMS works as the common authentication and authorization service for ISEE and ISEC. AUMS also works as the log server of ISEE and ISEC. AUMS gives security analysts anonymized data with permission administrators of ISEE and ISEC, and gets rules for generating suggestions from the security analysts. AUMS creates suggestions from monitored user

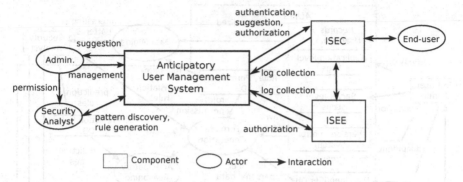

Fig. 3. Relationship among an anticipatory user management system, ISEE, and ISEC

status and the rules, and then sends the suggestions to end-users via ISEC. AUMS also gives the administrators suggestions.

Core of AUMS is suggestion generating mechanism. Figure 4 shows data flow diagram of the suggestion generating mechanism. The suggestion generating mechanism can be classified into two processes. One is a process to get prediction rules, best practices, anti-patterns, and suggestion rules from records of user behavior by doing data mining. Other is a process to generate suggestions from observed current status of an end-user, and already given prediction rules, best practices, anti-patterns, and suggestion rules.

When an end-user or an administrator of ISEC or ISEE does his/her tasks on ISEC, the one's behavior is recorded on AUMS via ISEC or ISEE. Data of user behavior can be classified into application-level logs, server-level logs, and changes of documents. Application-level logs are records of services of ISEC and ISEE that are used by end-users or administrators, i.e., who uses which service of ISEC or tool integrated in ISEE when. Server-level logs are logs of web servers, data management systems, mail servers, and other servers that are parts of ISEC or ISEE. Server-level logs denote user behavior inside of each service of ISEC or tool in ISEE, i.e., who accesses to which web page when, or who do what to which document when, or who will receive what e-mail. Changes of documents are managed by a version control system, like subversion [1], git [17], etc. Those data are used for accounting or auditing, i.e., monitoring current or past user status, and for discovering prediction rules, best practices, anti-patterns, and suggestion rules.

Security analysts discover prediction rules, best practices, anti-patterns, and suggestion rules from anonymized data extracted from AUMS by data mining. A prediction rule is a frequent sequential pattern among actions that users do on the ISEC or ISEE, i.e., which action is performed after doing which action. A best practice is a set of actions that are recommended. An anti-pattern is a set of actions that cause defects that become vulnerability or faults. A suggestion rule is a set of instructions to achieve a best practice or to avoid an anti-pattern. After discovering prediction rules, best practices, anti-patterns, and suggestion rules, the security analysts store them into Prediction rule DB, Behavior DB,

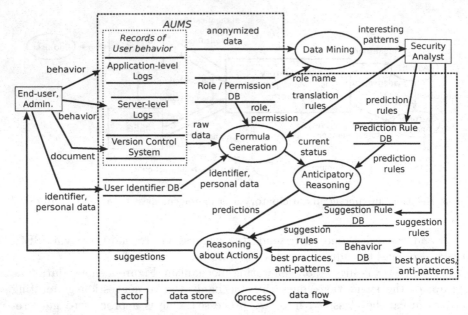

Fig. 4. Data-flow diagram of suggestion generating mechanism

and Suggestion rule DB. This discovery process deals with a set of records of user behavior statistically, so user identifiers in the records are replaced with temporary identifiers or role names.

The process to generate suggestions starts from getting current user status. Current user status is a set of observed facts. An observed fact is which action a user has done or what change the user has made. Current user status is extracted from records of user behavior. The current user status should be transformed into a set of logical formulas because anticipatory reasoning and reasoning about actions, which are introduced in section 3.2, are logic-based reasoning. Which data should be transformed into logical formulas and how to transform the data depend on what prediction rules, best practices, anti-patterns, and suggestion rules are used. Thus, security analysts provide AUMS with the instructions to transform which data into logical formulas and how to do it as transformation rules. After getting logical formulas that represent current user status, predictions are reasoned out from the current status and prediction rules by anticipatory reasoning. A prediction is a sentence that represents who do which action in the future. Finally, suggestions are reasoned out from deduced predictions, stored suggestion rules, best-practices, anti-patterns by reasoning about actions.

AUMS satisfies all requirements as mentioned in section 4.1. AUMS provides both ISEE and ISEC with facilities of AAA for satisfying R1. AUMS adopts anticipatory reasoning and reasoning about actions as mechanism of generating suggestions to satisfy R2, R4, and R5. Both anticipatory reasoning and reasoning about actions are domain independent. According to prediction rules, best practices, anti-patterns, and suggestion rules given by security analysts, AUMS can generate suggestions for design, development, management, and maintenance in individual-level, team-level, organization-level,inter-organization-level anticipatorily and actively. To satisfy R3: secure preservation of information related with users, all information related with users are gathered into AUMS.

4.3 Implementation Issues

Now, an anticipatory user management system (AUMS) is just a theoretical work. There are several issues to implement AUMS: how to implement anticipatory reasoning engine and action reasoning engine, how to effectively manage records of user behavior, how to discover prediction rules, best practices, anti-patterns, and suggestion rules.

Anticipatory reasoning engine and action reasoning engine can be implemented by using FreeEnCal [12, 18, 19, 41]. FreeEnCal is a forward reasoning engine with general-purpose, and can interpret specifications written in the formal language such that any user can use the formal language to describe and represent formulas and inference rules for deductive, simple inductive, and simple abductive reasoning. It also can reason out all or a part of logical theorem schemata of a logic system under the control conditions attached to the reasoning task specified by users, and all or a part of empirical theorems of a formal theory and facts under the control conditions attached to the reasoning task specified by users. Although the prototype of them were already implemented [8, 20, 38, 40, 46], the prototypes are not based on FreeEnCal. Therefore, their efficiency and generality is not enough.

Although the number of records of user behavior may become huge, monitoring current user status should be done as fast as possible. It is a contradictional requirement, but MapReduce [13] or Hadoop [60] can satisfy the requirement. MapReduce has been used by Google to generate scalable applications. MapReduce provides an interface that allows distributed computing and parallelization on clusters of computers. Hadoop is an open source version of MapReduce. Disadvantage of MapReduce and Hadoop is both of them cannot deal with update of stored data well. However, there is no problem because records of user behavior are used for only monitoring.

It is difficult to discover prediction rules, best practices, anti-patterns, and suggestion rules from records of user behavior. To establish a systematic method to discover them is a challenging issue in information security engineering. Many data mining methods have been proposed and developing [23, 53]. Those data mining methods are helpful for the discovery.

5 Concluding Remarks

In this paper, we have presented an anticipatory user management system as a user management mechanism in Information Security Engineering Environment (ISEE) and Information Security Engineering Cluster (ISEC). An anticipatory user management system is an information system that provides ISEE and ISEC with anticipatory user management. Anticipatory user management is to manage user identifiers, user profiles, permission/roles, and records of user behavior, and to control not only access to stored data and resources of a target information system, but also user behavior by giving appropriate suggestions anticipatorily and actively. Anticipatory user management is a new type of user management for SaaS-based cloud services with facility to control user behavior.

This work is just a theoretical work. There are several implementation issues: how to implement anticipatory reasoning engine and action reasoning engine, how to effectively manage records of user behavior, how to discover prediction rules, best practices, anti-patterns, and suggestion rules.

References

1. Apache Software Foundation: Apache Subversion, http://subversion.apache.org/ (accessed at February 19, 2014)
2. Bao, D., Miura, J., Zhang, N., Goto, Y., Cheng, J.: Supporting Verification and Validation of Security Targets with ISO/IEC 15408. In: The 2nd International Conference on Mechatronic Sciences, Electric Engineering and Computer (MEC 2013), Shenyang, China, pp. 2621–2628. IEEE (2013)
3. Brown, W.J., Malveau, R.C., McCormick, H.W., Mowbray, T.J.: AntiPatterns: Refactoring Software, Architectures, and Projects in Crisis. John Wiley & Sons, Inc. (1998)
4. Chen, B., Chi, X., Wu, H.: A Model for User Management in Grid Computing Environments. In: Li, M., Sun, X., Deng, Q., Ni, J. (eds.) GCC 2003, Part I. LNCS, vol. 3032, pp. 732–737. Springer, Heidelberg (2004)
5. Cheng, J.: Anticipatory Reasoning-Reacting Systems. In: International Conference on Systems, Development and Self-organization, Beijing, China, pp. 161–165 (2002)
6. Cheng, J.: Temporal Relevant Logic as the Logical Basis of Anticipatory Reasoning-Reacting Systems. In: Dubois, D.M. (ed.) Computing Anticipatory Systems: CASYS 2003 - Sixth International Conference, Liege, Belgium, August 11-16. AIP Conference Proceedings, vol. 718, pp. 362–375. American Institute of Physics (2004)
7. Cheng, J.: Temporal Deontic Relevant Logic as the Logical Basis for Decision Making Based on Anticipatory Reasoning. In: 2006 IEEE International Conference on Systems, Man and Cybernetics, Taipei, Taiwan, pp. 1036–1041. IEEE Systems, Man, and Cybernetics Society (2006)
8. Cheng, J.: Adaptive Prediction by Anticipatory Reasoning Based on Temporal Relevant Logic. In: 2008 Eighth International Conference on Hybrid Intelligent Systems, Barcelona, Spain, pp. 410–416. IEEE (2008)
9. Cheng, J., Goto, Y., Horie, D.: ISEE: An Information Security Engineering Environment. In: International Conference on Security and Cryptography, Milan, Italy, pp. 395–400. INSTICC Press (2009)
10. Cheng, J., Goto, Y., Kitajima, N.: Anticipatory Reasoning about Mobile Objects in Anticipatory Reasoning-Reacting Systems. In: Dubois, D.M. (ed.) Computing Anticipatory Systems: CASYS 2007 - Eighth International Conference. AIP Conference Proceedings, vol. 718, pp. 244–254. American Institute of Physics (2008)

11. Cheng, J., Goto, Y., Morimoto, S., Horie, D.: A Security Engineering Environment Based on ISO/IEC Standards: Providing Standard, Formal, and Consistent Supports for Design, Development, Operation, and Maintenance of Secure Information Systems. In: The 2nd International Conference on Information Security and Assurance, Busan, Korea, pp. 350–354. IEEE Computer Society (2008)
12. Cheng, J., Nara, S., Goto, Y.: FreeEnCal: A Forward Reasoning Engine with General-Purpose. In: Apolloni, B., Howlett, R.J., Jain, L. (eds.) KES 2007, Part II. LNCS (LNAI), vol. 4693, pp. 444–452. Springer, Heidelberg (2007)
13. Dean, J., Ghemawat, S.: MapReduce: Simplified Data Processing on Large Clusters. Communications of the ACM 51(1), 107–113 (2008)
14. Denemark, J., Jankowski, M., Křenek, A., Matyska, L., Meyer, N., Ruda, M., Wolniewicz, P.: Best Practices of User Account Management with Virtual Organization Based Access to Grid. In: Wyrzykowski, R., Dongarra, J., Meyer, N., Waśniewski, J. (eds.) PPAM 2005. LNCS, vol. 3911, pp. 633–642. Springer, Heidelberg (2006)
15. Dubois, D.M.: Computing Anticipatory Systems with Incursion and Hyperincursion. In: Dubois, D.M. (ed.) The First International Conference on Computing Anticipatory Systems. AIP Conference Proceedings, vol. 437, pp. 3–30. American Institute of Physics (1998)
16. Dubois, D.M.: Introduction to Computing Anticipatory Systems. International Journal of Computing Anticipatory Systems 2, 3–14 (1998)
17. Git: Git, http://git-scm.com/ (accessed at February19, 2014)
18. Goto, Y., Gao, H., Tsuji, T., Cheng, J.: Practical Usage of FreeEnCal: An Automated Forward Reasoning Engine for General-Purpose. In: The International Conference on Machine Learning and Cybernetics, Xi'an, China, pp. 1878–1883. IEEE (2012)
19. Goto, Y., Koh, T., Cheng, J.: A General Forward Reasoning Algorithm for Various Logic Systems with Different Formalizations. In: Lovrek, I., Howlett, R.J., Jain, L.C. (eds.) KES 2008, Part II. LNCS (LNAI), vol. 5178, pp. 526–535. Springer, Heidelberg (2008)
20. Goto, Y., Nara, S., Cheng, J.: Efficient Anticipatory Reasoning for Anticipatory Systems with Requirements of High Reliability and High Security. International Journal of Computing Anticipatory Systems 14, 156–171 (2004)
21. Grzonkowski, S., Gzella, A., Krawczyk, H., Kruk, S.R., MartinRecuerda, F., Woroniecki, T.: D-FOAF - Security Aspects in Distributed User Management System. In: The IEEE International Conference on Technologies for Homeland Security and Safety (2005)
22. Hacker, T.J., Athey, B.D.: A Methodology for Account Management in Grid Computing Environments. In: Lee, C.A. (ed.) GRID 2001. LNCS, vol. 2242, pp. 133–144. Springer, Heidelberg (2001)
23. Han, J., Kamber, M., Pei, J.: Data Mining: Concepts and Techniques, 3rd edn. Morgan Kaufmann Publishers (2011)
24. Hogan, M., Sokol, A.: NIST Cloud Computing Standards Roadmap, Version 2 (2013)
25. Horie, D., Goto, Y., Cheng, J.: Development of ISEE: An Information Security Engineering Environment. In: 2009 Second International Symposium on Electronic Commerce and Security, Nanchang, China, pp. 338–342. IEEE (2009)
26. Horie, D., Kasahara, T., Goto, Y., Cheng, J.: A New Model of Software Life Cycle Processes for Consistent Design, Development, Management, and Maintenance of Secure Information Systems. In: 2009 Eighth IEEE/ACIS International Conference on Computer and Information Science, Shanghai, China, pp. 897–902. EEE Computer Society (2009)

27. Horie, D., Morimoto, S., Azimah, N., Goto, Y., Cheng, J.: ISEDS: An Information Security Engineering Database System Based on ISO Standards. In: 2008 Third International Conference on Availability, Reliability and Security, Barcelona, Spain, pp. 1219–1225. IEEE Computer Society (2008)
28. Horie, D., Yajima, K., Azimah, N., Goto, Y., Cheng, J.: GEST: A Generator of ISO/IEC 15408 Security Target Templates. In: Lee, R., Hu, G., Miao, H. (eds.) Computer and Information Science 2009. SCI, vol. 208, pp. 149–158. Springer, Heidelberg (2009)
29. International Organization for Standardization: ISO/IEC 15408:2009, Information Technology – Security Techniques - Evaluation Criteria for IT Security (2009)
30. International Organization for Standardization: ISO/IEC 18045:2008 Information Technology – Security Techniques – Methodology for IT Security Evaluation (2008)
31. International Organization for Standardization: ISO/IEC 27000: Information Technology – Security Techniques – Information Security Management Systems – Overview and Vocabulary (2009)
32. International Organization for Standardization: ISO/IEC 27001: Information Technology – Security Techniques – Information Security Management Systems – Requirements (2005)
33. International Organization for Standardization: ISO/IEC 27002: Information Technology – Security Techniques – Code of Practice for Information Security Management (2005)
34. International Organization for Standardization: ISO/IEC 27003: Information Technology – Security Techniques – Information Security Management Implementation Guidance (2010)
35. International Organization for Standardization: ISO/IEC 27004: Information Technology – Security Techniques – Information Security Management – Measurement (2009)
36. International Organization for Standardization: ISO/IEC 27005: Information Technology – Security Techniques – Information Security Risk Management (2011)
37. Iqbal, A., Horie, D., Goto, Y., Cheng, J.: A Database System for Effective Utilization of ISO/IEC 27002. In: 2009 Fourth International Conference on Frontier of Computer Science and Technology, Shanghai, China, pp. 607–612. IEEE Computer Society (2009)
38. Kitajima, N., Goto, Y., Cheng, J.: Fast Qualitative Reasoning about Actions for Computing Anticipatory Systems. In: 2008 Third International Conference on Availability, Reliability and Security, Barcelona, Spain, pp. 171–178. IEEE Computer Society (2008)
39. Kitajima, N., Goto, Y., Cheng, J.: Development of a Decision-Maker in an Anticipatory Reasoning-Reacting System for Terminal Radar Control. In: Corchado, E., Wu, X., Oja, E., Herrero, Á., Baruque, B. (eds.) HAIS 2009. LNCS, vol. 5572, pp. 68–76. Springer, Heidelberg (2009)
40. Kitajima, N., Nara, S., Goto, Y., Cheng, J.: A Deontic Relevant Logic Approach to Reasoning. International Journal of Computing Anticipatory Systems 20, 177–190 (2008)
41. Koh, T., Goto, Y., Cheng, J.: A Fast Duplication Checking Algorithm for Forward Reasoning Engines. In: Lovrek, I., Howlett, R.J., Jain, L.C. (eds.) KES 2008, Part II. LNCS (LNAI), vol. 5178, pp. 499–507. Springer, Heidelberg (2008)
42. Li, J., Zhang, C.: A Three-Dimensional Role Based User Management Model in Web Information Systems. In: Lu, W., Cai, G., Liu, W., Xing, W. (eds.) Proceedings of the 2012 International Conference on Information Technology and Software Engineering. LNEE, vol. 210, pp. 657–665. Springer, Heidelberg (2013)

43. Lin, J., Lu, X., Yu, L., Zou, Y., Zha, L.: VegaWarden: A Uniform User Management System for Cloud Applications. In: 2010 IEEE Fifth International Conference on Networking, Architecture, and Storage, Macau, China, pp. 457–464. IEEE (2010)
44. Liu, F., Dogdu, E.: A User Management System for Federated Databases Using Web Services. In: Gavrilova, M.L., et al. (eds.) ICCSA 2006. LNCS, vol. 3983, pp. 88–97. Springer, Heidelberg (2006)
45. Liu, L.M., Xu, Z., Li, W.: A Layered Grid User Expression Model in Grid User Management 2 The Grid User Expression RUS Model. In: Li, M., Sun, X.H., Deng, Q.N., Ni, J. (eds.) GCC 2003. LNCS, vol. 3033, pp. 1055–1058. Springer, Heidelberg (2004)
46. Nara, S., Shang, F., Omi, T., Goto, Y., Cheng, J.: An Anticipatory Reasoning Engine for Anticipatory Reasoning-Reacting Systems. International Journal of Computing Anticipatory Systems 18, 225–234 (2006)
47. Rosen, R.: Anticipatory Systems – Philosophical, Mathematical and Methodological Foundations. Pergamon Press (1985)
48. Sandhu, R.S., Feinstein, H.L., Youman, C.E.: Role Based Access Control Models. Computer 29(2), 38–47 (1996)
49. Suhaimi, A.I.H., Goto, Y., Cheng, J.: An Engineering Environment for Supporting Information Security Management Systems. In: Kim, T.H., et al. (eds.) SecTech, CA, CES3 2012. CCIS, vol. 339, pp. 30–37. Springer, Heidelberg (2012)
50. Suhaimi, A.I.H., Goto, Y., Cheng, J.: An Analysis of Software Supportable Tasks in Information Security Management System Life Cycle Processes. In: International Conference on Information and Social Science, Nagoya, Japan, pp. 29–58 (2013)
51. Suhaimi, A.I.H., Manji, T., Goto, Y., Cheng, J.: A Systematic Management Method of ISO Information Security Standards for Information Security Engineering Environments. In: Abd Manaf, A., Zeki, A., Zamani, M., Chuprat, S., El-Qawasmeh, E. (eds.) ICIEIS 2011, Part I. CCIS, vol. 251, pp. 370–384. Springer, Heidelberg (2011)
52. Sun, G., Yajima, K., Miura, J., Goto, Y., Cheng, J.: A Supporting Tool for Creating and Maintaining Security Targets according to ISO/IEC 15408. In: 2012 IEEE International Conference on Computer Science and Automation Engineering, Beijing, China, pp. 745–749. IEEE (2012)
53. Wu, X., Kumar, V., Ross Quinlan, J., Ghosh, J., Yang, Q., Motoda, H., McLachlan, G.J., Ng, A., Liu, B., Yu, P.S., Zhou, Z.H., Steinbach, M., Hand, D.J.: Top 10 Algorithms in Data Mining. Knowledge and Information Systems 14(1), 1–37 (2008)
54. Xu, L., Shi, K., Goto, Y., Cheng, J.: ISEC: An Information Security Engineering Cloud. In: 2012 IEEE International Conference on Computer Science and Automation Engineering, Beijing, China, pp. 750–753. IEEE (2012)
55. Xu, L., Wang, B., Zhang, N., Goto, Y., Cheng, J.: Providing Users with Suitable Services of Information Security Engineering Cloud based on ISO/IEC 15408. In: 2013 IEEE 4th International Conference on Software Engineering and Service Science, Beijing, China, pp. 321–325. IEEE (2013)
56. Yajima, K., Morimoto, S., Horie, D., Azreen, N.S., Goto, Y., Cheng, J.: FORVEST: A Support Tool for Formal Verification of Security Specifications with ISO/IEC 15408. In: 2009 International Conference on Availability, Reliability and Security, Fukuoka, Japan, pp. 624–629. IEEE Computer Society (2009)
57. Zhang, N., Bao, D., Xu, L., Suhaimi, A.I.H., Miura, J., Goto, Y., Cheng, J.: Supporting Tools for Software Supportable Tasks Related with ISO/IEC 15408. In: The 2nd International Conference on Mechatronic Sciences, Electric Engineering and Computer, Shenyang, China, pp. 2002–2006. IEEE (2013)

58. Zhang, N., Suhaimi, A.I.H., Goto, Y., Cheng, J.: An Analysis of Software Support-able Tasks Related with ISO/IEC 15408. In: The 9th International Conference on Computational Intelligence and Security, E'Mei Shan, China, pp. 601–606. IEEE Computer Society (2013)
59. Zhu, L., Kent, R.D., Aggarwal, A., Viranthi, P., Rahman, Q., Elamsy, T., Ejelike, O., Statement, P.: Construction of a Webportal and User Management Framework for Grid. In: 21st International Symposium on High Performance Computing Systems and Applications, Saskatoon, SK, pp. 3–9. IEEE Computer Society (2007)
60. Zikopoulos, P., Eaton, C.: Understanding Big Data: Analytics for Enterprise Class Hadoop and Streaming Data. McGraw-Hill Osborne Media (2011)

Active and Personalized Services
in an Information Security Engineering Cloud
Based on ISO/IEC 15408

Liqing Xu, Yuichi Goto, Ahmad Iqbal Hakim Suhaimi,
Ning Zhang, and Jingde Cheng

Department of Information and Computer Science,
Saitama University, Saitama, 338-8570, Japan
{xuliqing,gotoh,iqbal,zhang,cheng}@aise.ics.saitama-u.ac.jp

Abstract. An Information Security Engineering Environment (ISEE) based on ISO/IEC security standards has been proposed. It integrates various tools such that its users can use these tools to ensure the whole security of their target information system at anytime consistently and continuously according to ISO/IEC security standards. But ISEE can only provide its services passively, i.e., when users use ISEE, they have to give some commands or instructions to ISEE. Because crackers are active persons who can get knowledge and skills day after day and then continuously attack the weakest point or connection in each target system always with new techniques, some active services and personalized services to defend attacks and prevent damage beforehand are very desirable to various users of ISEE. We have proposed an Information Security Engineering Cloud (ISEC) as a platform to provide various active services and personalized services based on ISEE to its various users in a way of cloud services. ISO/IEC 15408, as one of the most important international standards, plays an important role to ensure the whole security of target information/software systems, and therefore, has been adopted as the core standard in ISEC. This paper presents a control mechanism to provide active and personalized serviced based on ISO/IEC 15408. In order to realize this mechanism, we defined active and personalized services of ISEC, and analyzed necessary data of checkpoints, which are the items controlled by a series of tasks for managing task progress based on ISO/IEC 15408. Based on the analysis, we show how to provide active and personalized services to meet the different needs of various users.

Keywords: Information Security Engineering Cloud, Active Services, Personalized Services.

1 Introduction

An Information Security Engineering Environment (ISEE) [3–5] based on ISO/IEC security standards has been proposed. It integrates various tools such that its users can use these tools to ensure the whole security of their target information system at anytime consistently and continuously according to ISO/IEC

M. Chau et al. (Eds.): PAISI 2014, LNCS 8440, pp. 35–48, 2014.

security standards. But ISEE can only provide its services passively, i.e., when users use ISEE, they have to give some commands or instructions to ISEE. Even though it provides integrates various tools and comprehensive facilities, it cannot support anticipatory services that its users need.

In order to provide its users with comprehensive support, active and personalized services are indispensable in ISEE. Because crackers are active persons who can get knowledge and skills day after day and then continuously attack the weakest point or connection in each target system always with new techniques, some active services and personalized services to defend attacks and prevent damage beforehand are very desirable to various users of ISEE.

Therefore, we have proposed an Information Security Engineering Cloud (ISEC) [13] as a platform to provide various active services and personalized services based on ISEE to its various users in a way of cloud services. ISEC is a cloud service has not only all functions of ISEE, but also can provide active and personalized services to designers, developers, administrators/end-users, and maintainers of target systems. It anticipates necessary services that users need, provides various active and personalized services, as well as manage these services. ISEC provides services for its users based on their roles/responsibilities to perform all tasks in correct sequence, without any lack, and satisfying a certain level of quality of the whole security of the systems. ISO/IEC 15408 [10], as one of the most important international standards for information security, has been adopted as the core standard in ISEC. It plays an important role to ensure the whole security of target information/software systems.

In order to defend attacks and prevent damage beforehand, ISEC should have a control mechanism to provide active and personalized services to various users. However, there is such method has existed until now.

This paper presents a control mechanism to provide active and personalized services based on ISO/IEC 15408. In order to realize this mechanism, we defined active and personalized services of ISEC, and analyzed necessary data of checkpoints, which are the items controlled by a series of tasks for managing task progress based on ISO/IEC 15408. Based on the analysis, we show how to provide active and personalized services to meet the different needs of various users.

The rest of this paper is organized as follows. Section 2 introduces ISEC. Section 3 shows active service and personalized service in ISEC. Section 4 presents analysis of progress of tasks based on ISO/IEC 15408. Section 5 presents a mechanism of active service and personalized service. Section 6 presents a use case. Some concluding remarks are given in Section 7.

2   An Information Security Engineering Cloud

ISEC should provide various services to its users in a way of cloud services. From the viewpoint of security engineering, all tasks related with security functions of a target information system are important, and should be performed in an appropriate sequence, without any lack, and satisfying a certain level of quality. Therefore, it is important to control the access to the tools of ISEE that provide services for the tasks. However, controlling such access from various kinds of users is difficult if the tools are not provided in a way of cloud services. Thus, it is demanded that ISEC should be as a cloud service to provide services for supporting the tasks.

Figure 1 shows a relationship between ISEE and ISEC. ISEE processes the actual tasks by collaboration of various tools, a central database system, and a work flow executor. ISEC retrieves requests from users, sends them to ISEE, and in return, gives appropriate suggestions to the users by a suggestion generator, a common interface, and a work flow generator.

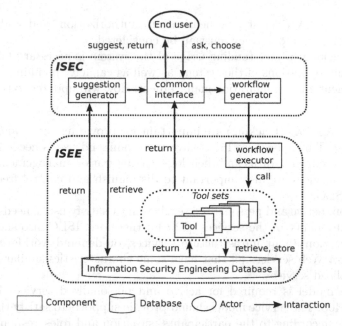

Fig. 1. Relationship between ISEE and ISEC

There are many services and various participants in ISEC based on ISO/IEC 15408. Services in ISEC support 30 processes, 79 tasks, and 94 actions, and are actively and personally provided to 15 kinds of participants. Users of ISEC are designers, developers, maintainers, and administrators who establish and manage ISEC and its services, and the participants who use the services. ISEC supports its users to find appropriate services according to their roles/responsibilities. ISEC not only can provide all general services, but also can provide specific and desirable services to satisfy the users' needs according to their situations.

3 Active Services and Personalized Services in ISEC

The major points in ISEC are active services and personalized services. Active service of ISEC is an advanced service which provide active capabilities of ISEC by predicting users' needs in advance and actively providing needed service to various users. Personalized service of ISEC is a private customized service which is based on the understanding of user's needs, and provides satisfactory and efficient service to each user with customized service. Active and personalized services can enable effective utilization of ISO/IEC 15408 to provide comprehensive facilities for design, development, operation, and maintenance of target information/software systems with high reliability and security requirements.

In order to provide cloud services, it is necessary for ISEC to consider the following items [2].

- Security: IAAA (identity, authentication, authorization, and audit), data confidentiality, and privacy must all be considered.
- Usage tracking: Understanding usage of the system is necessary for a minimum daily operations of the system, as well as capacity planning.
- Deployment and configuration management: How will updates to the system be deployed.

IAAA and usage logs (tracking) are important for providing active and personalized service. This is because ISEC have to recommend user's needed services based on their current status and their logs. Configuration management and deployment of services are also important to distinguish and control free or paid services of ISEC.

In addition, active and personalized services must satisfy users' needs in order to ensure the quality of the services. [12]. Furthermore, ISEC also should take into consideration about infrastructure services, on-demand workforce service [6], and Alexa Web services [6] that have some characteristics similar to active and personalized services.

A service model is required for active and personalized services based on ISO/IEC 15408. The service model should provide supports to satisfy the participants' needs according to the participants' situation and roles/responsibilities. This is because ISEC should provide many services supporting tasks based on ISO/IEC 15408 to various participants. However, no such model has existed until now.

Therefore, we present a model based on a degree of personalization to provide active and personalized services. Figure 2 shows the service model for the degree of personalization. First layer is non-activeness and non-personalization. Second layer is non-activeness and personalization that helps to provide appropriate services according to the participant's role. Third layer is activeness and personalization that helps to inform the users and enforces related services to the users.

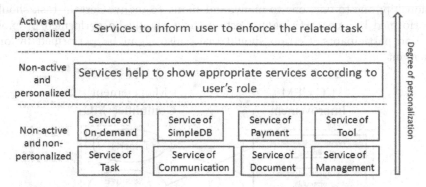

Fig. 2. Service model for active and personalized of ISEC

Cloud-computing service models are often self-service, even in internal models [8]. We will develop application, provide an execution platform, and run the model actively and personalization. Application are important to the cloud computing stack [1] and they provide end users with an integrated service.

4 Analysis of Progress of Tasks Based on ISO/IEC 15408

4.1 Overview

In order to provide active and personalized services based ISO/IEC 15408, a control mechanism that can consider user's situations based on their roles/responsibilities and progress of related task is necessary. Therefore, it is needed to analysis progress of tasks based on ISO/IEC 15408.

We have analyzed the relationships among tasks, documents, and participants based on ISO/IEC 15408. There has been an analysis [15] about documents, processes, and participants based on ISO/IEC 15408 as well as their relationship. However, the analysis is not detailed in task level as one process consists of several tasks. Because providing active and personalized services should be based on participants' roles/responsibilities and progress of tasks, it is important to clarify the relationship among the tasks in ISO/IEC 15408. Thus, we made further analysis on the relationship among documents, tasks, and participants. Figure 3 shows a part of relationship between participants and tasks related with ISO/IEC 15408. The relationship is based on roles/responsibilities of participants. The meanings for abbreviations used in the figure are as follows. CC means The Common Criteria for Information Technology Security Evaluation, CEM means Common Methodology for Information Technology Security Evaluation, CCMB means Common Criteria Maintenance Board, and CCDB means Common Criteria Development Board.

From the figure, the tasks are related with particpants in various kinds of relationship: one-to-one, one-to-many, and many-to-many. Certain task should be performed by one participant. Certain tasks should be performed by more than one participants. Certain should tasks also shoul be performed by one participant.

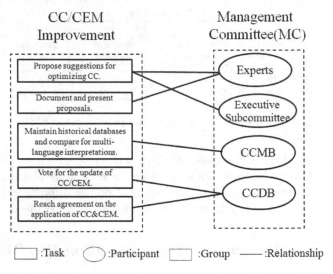

Fig. 3. A part of relationship between participants and tasks related with ISO/IEC 15408

In order to realize active services and personalized services, a control mechanism should provide different services to various participants based on their roles/responsibilities and progress of tasks in ISO/IEC 15408. There a total of five processes based on ISO/IEC 15408: development, preparation, evaluation, certification, and consumption. We have analyzed relationship among participants, tasks, and documents in the five processes. The analysis results are shown in Figure 4, Figure 5, Figure 6, Figure 7 respectively. The figures show the sequence of tasks and participants for the tasks. The meanings for abbreviations used in the figures are as follows. ES means Executive Subcommittee. CB is a short form for Certification/Validation Body. CP is a short form for Change Proposals. AP is short for Assurance Packages. TC is a short form for Technical Consistency. OR is a short form for Observation Report. ETR is a short form for Evaluation Technical Report. CCC is a short form Common Criteria Certificates. CR is a short form for Certification Report. D-CPL is a short form Domestic Certified/Validated Products List. I-CPL is a short form for International Certified/Validated Products List.

Furthermore, the control mechanism should manage progress of tasks based on ISO/IEC 15408. The main reason is because the there is a lot of tasks in the standards and there are related each other in complex relationship.

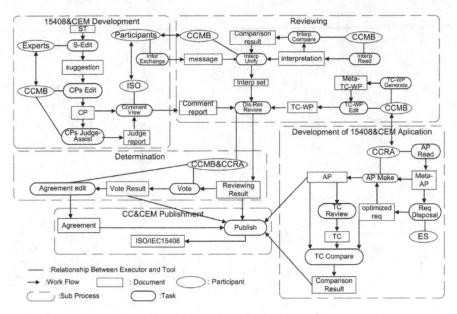

Fig. 4. Relationship of tasks, documents, and participants in development process related with ISO/IEC 15408

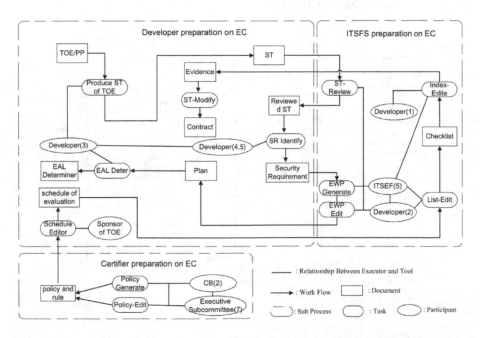

Fig. 5. Relationship of tasks, documents, and participants in preparation process related with ISO/IEC 15408

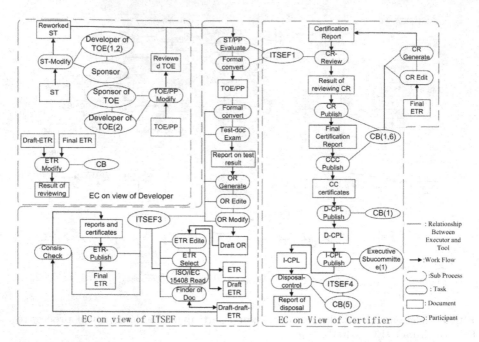

Fig. 6. Relationship of tasks, documents, and participants in evaluation and certification process related with ISO/IEC 15408

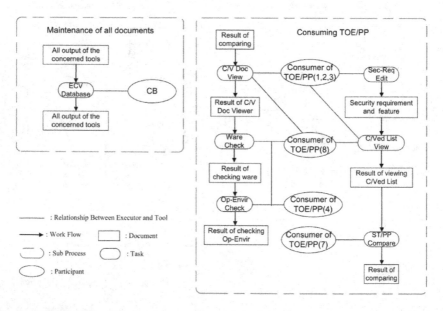

Fig. 7. Relationship of tasks, documents, and participants in consuming process related with ISO/IEC 15408

4.2 Checkpoints

Checkpoints are items used to manage progress of tasks of ISO/IEC 15408. We clarified checkpoints for each task in corresponding process. It is important to clarify the checkpoints to check when the associated documents are generated and its should be checked at the start of each task. Thus, the checkpoint is allocated in the beginning of new task execution and new document generation. In detail, checkpoint is an item that uses to check action of the task. In order to complete the tasks, participants have to perform all actions related to the task. Because all tasks should be performed in an appropriate sequence, it is necessary to control and check each action. In addition, checkpoint time also necessary to check the time when the task is started, the time when the stored contents of the document will be generated, and the time when the task is ended. Before starting a new task, the related participant and document should be prepared in advanced. When the task start, new documents are created. The progress of these documents should be created and managed for later checking.

According to the analysis results of the tasks, we have enumerated 681 checkpoints. Based on the similarity of the checkpoints of each task, we summarized 12 common checkpoints. The 12 common checkpoints are as follows.

1. To check whether or not the document has been prepared (A).
2. To check whether or not the document has been called.
3. To check whether or not the document called time has been recorded.
4. To check whether or not participants has logged in to the system (B).
5. To investigate how many participants.
6. To check whether or not new documents have been created (C).
7. To check whether or not the new document created time has been recorded.
8. To investigate the percentage of the new created contents of document (D).
9. To check whether or not the time of the completely created document has been recorded (E).
10. To check whether the time of communication for participants exist or not (F).
11. To check whether the document has dependency or not (G).
12. To check the necessary for participants agreement.

Data from the checkpoint are important for algorithm of active and personalized services. After clarifying checkpoints of tasks, we collect data of each checkpoint to manage progress of tasks. From above checkpoints, if the checkpoint is "false", the data of the checkpoint is represented by "0", if the checkpoint is "true", the data of the checkpoint is represented by "1". Among the 12 common checkpoints, there are 7 necessary checkpoints (A - G) which all data of the checkpoints have to be obtained at the same time in order to provide active and personalized services.

4.3 An Algorithm

An algorithm is needed to process the obtained data from the checkpoints to judge status of tasks. ISEC provides the participants with the different services

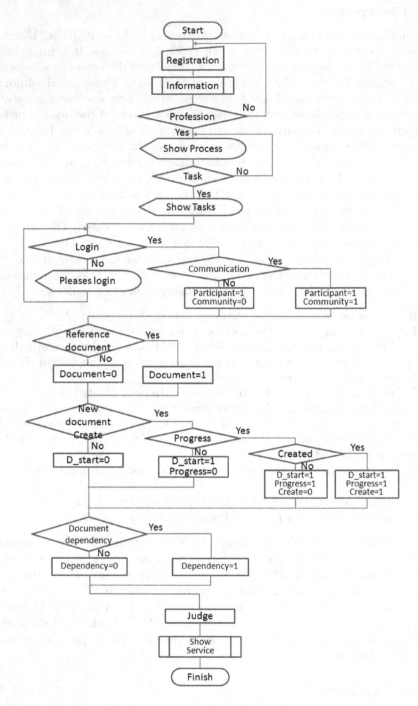

Fig. 8. An algorithm to obtain data

on the basis of the algorithm depending on the progress of the tasks. In addition, dependency of documents and status of documents are important for the algorithm. If such information does not exist, data from the checkpoint cannot be obtained, and therefore status of the tasks cannot be judged. Thus, the algorithm must clarify relationship among document, task, and participant as well as dependency of document related to the task.

We have proposed an algorithm to judge status of tasks. Figure 8 illustrates a flowchart of the algorithm. First, based on registered information the algorithm judges related processes or tasks for the participants. Second, the algorithm checks whether the participant has logged or not. Third, the algorithm checks whether the time of communication for participant exists or not. Fourth, the algorithm checks whether documents have been referred or not. Fifth, the algorithm checks whether creation of a new document has been started or not. If the document has been started, it needs to check the progress of the created document and completion time of the document. Sixth, the algorithm checks whether a new document has dependency relation or not. Finally, the algorithm judges what services need to be provided based on status of the status of tasks.

5   A Control Mechanism of Active and Personalized Services

A control mechanism of active and personalized services is a mechanism to control access authority, progress of tasks, and services for active services and personalized services. The control mechanism is shown in figure 9. Access authority is controlled to give participants access authority according to their roles/responsibilities. Tasks are controlled to control document status and action status of tasks. Services are controlled to provide general services [14], active services, and personalized services according to tasks status.

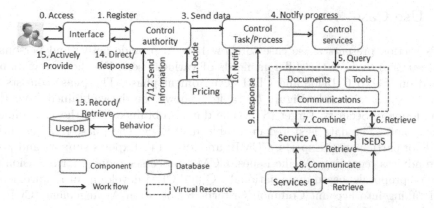

Fig. 9. A mechanism to control access authorities, tasks, and services

The mechanism consists of 9 components: interface, control authority, control task/process, control service, pricing, virtual resource, service, UserDB, and ISEDS [9]. Interface component supports to provide participants with required services through a web interface. Control authority component supports to give participant access authority based on information of registration and logs. Control task/process component supports to manage tasks and record status of the tasks. Control service component supports to manage participant's services by inferring appropriate service to the participants according to the progress of task. Behavior component supports to manage information of registration and logs. Pricing component supports to determine whether certain services are for free or need payment, set time duration to use free services, and set price for paid services. Virtual resource component supports to provide needed resource for documents, tools, and communications. Service component supports to combine services based on inference component. UserDB manages information of registration and logs. ISEDS is a database system that manages data of ISO/IEC security standards and their related documents, data of published cases, and data acquired by participants in their tasks.

The mechanism not only can judge suitable services according to participants' roles/responsibilities and behaviors, but also can provide participants with active and personalized services according to progress of tasks/processes. ISEC provides participants with active services based on their behavior, and provides personalized services are based on progress of tasks and participant's roles/ responsibilities. The mechanism uses neighborhood methods [11] to recommend participant with personalized services. Neighborhood methods are centered on computing the relationships between items which in this case are relationships between participants. Behavior component, control authority component, and control task/process component are the most important components in order to provide participants with active and personalized services.

6   Use Case

This section presents a use case to show how to provide active and personalized services to meet the different needs of various users. We show an example based on ISO/IEC 15408 and CEM development phase. The phase consists of three tasks as shown in Figure 4. Table 1 shows the data obtained from the production of Security Target (ST). The data is obtained from the 12 common checkpoints. The data is stored in a table in ISEDS. The phase requires three kinds of participants: Experts, CCMB, and ISO [14]. Experts support and provide advices regarding specific issues. CCMB processes request for inclusion of change proposals, based upon national CC and CEM development requirements and taking into account Common Criteria Recognition Arrangement (CCRA) requirements [7]. ISO ensures to develop ISO standards according to consensus, industry wide, voluntary standards. Items from A to G in Table 1 are the data 7 necessary checkpoints. Using the data in Table 1, the algorithm judges progress of tasks and then provides different services to different users. We list six services

Table 1. Relationship of obtained data and services

Case	A 1/0	B 1/0	C 1/0	D 1/0	E 1/0	F 1/0	G 1/0	Task status	Service No.
1	0	0	0	0	0	0	0	Before the implementation	① ②
2	1	0	0	0	0	0	0	Before the implementation	① ②
3	1	1	0	0	0	0	0	In progress	③ ④
4	1	1	1	0	0	0	0	In progress	③ ④
5	1	1	1	1	0	0	0	In progress	③ ④
6	1	1	1	1	1	0	0	In progress	③ ④
7	1	1	1	1	1	1	0	In progress	③ ④
8	1	1	1	1	1	1	1	After the implementation	⑤ ⑥ ⑦

in Table 1 as follows. All seven services belong to 94 services based on ISO/IEC 15408 [14].

1. ISEC informs experts to begin the production of ST.
2. ISEC gives experts access to use appropriate tools (ST-Generator, ST-Editor) to produce ST.
3. ISEC gives experts access to refer and retrieve logs of tasks related with ST production.
4. ISEC suggest experts to use appropriate tools (ST-comparer, ST-Modifier).
5. ISEC informs CCMB to generate new suggestion document.
6. ISEC informs CCMB to begin the CPs editing task.
7. ISEC suggest ISO to begin the commenting/viewing task.

From above, the control mechanism is capable to provide active and personalized services for various participants with different needs. The above use case showed the mechanism which is used to control access authority according to users' roles/responsibilities, and to control the services according to the progress of tasks. The use case also showed that different services are provided actively and personally to various participants according the task status. Task status is judge by algorithm based on data obtained from checkpoint to judge by the algorithm.

7 Concluding Remarks

This paper proposed a mechanism to control authorities, tasks, and services in order to provide active and personalized services in ISEC based on ISO/IEC 15408, and showed how to provide such services to meet the different needs of various users in order to help them to perform their tasks efficiently.

In the future, in order to manage the relationships among the processes, tasks, documents, as well as participants in ISEC, a systematic management method is necessary. ISEC encounters many issues to manage participants, processes, tasks, and documents systematically because there are many processes, tasks, documents, participants that are related to each other with complex relationships. Although we have clarified these relationships, it is not easy to manage the relationships. How to manage the relationships is still a remaining issue.

References

1. Buyya, R., Pandey, S., Vecchiola, C.: Cloudbus Toolkit for Market-Oriented Cloud Computing. In: Jaatun, M.G., Zhao, G., Rong, C. (eds.) CloudCom. LNCS, vol. 5931, pp. 24–44. Springer, Heidelberg (2009)
2. Cavage, M.: There's Just No Getting around It: You're Building a Distributed System. Int. J. ACM Queue 11(4), 80–89 (2013)
3. Cheng, J., Goto, Y., Morimoto, S., Horie, D.: A Security Engineering Environment Based on ISO/IEC Standards: Providing Standard, Formal, and Consistent Supports for Design, Development, Operation, and Maintenance of Secure Information Systems. In: Proceedings of the 2nd International Conference on Information Security and Assurance (ISA 2008), pp. 350–354. IEEE Computer Society Press (2008)
4. Cheng, J., Goto, Y., Horie, D.: ISEE: An Information Security Engineering Environment. In: Proceedings of the International Conference on Security and Cryptography (SECRYPT 2009), pp. 395–400. INSTICC Press (2009)
5. Cheng, J., Goto, Y., Horie, D., Miura, J., Kasahara, T., Iqbal, A.: Development of ISEE: An Information Security Engineering Environment. In: Proceedings of the 7th IEEE International Symposium on Parallel and Distributed Processing with Applications (ISPA 2009), pp. 505–510. IEEE Computer Society Press (2009)
6. Classmethod: List of Services Provided by the Amazon Cloud, http://classmethod.jp/solutions/aws/tech/ (accessed June 20, 2013)
7. Common Criteria Project: Common Criteria Portal, http://www.commoncriteriaportal.org/ (accessed June 20, 2013)
8. Creeger, M.: Cloud Computing: An Overview. Int. J. ACM Queue 7(5), 3–4 (2009)
9. Horie, D., Morimoto, S., Azimah, N., Goto, Y., Cheng, J.: ISEDS: An Information Security Engineering Database System Based on ISO Standards. In: Proceedings of the 3rd International Conference on Availability, Reliability and Security (ARES 2008), pp. 1219–1225. IEEE Press (2008)
10. International Organization for Standardization: ISO/IEC 15408:2009, Information Technology - Security Techniques - Evaluation Criteria for IT Security (2009)
11. Koren, Y., Bell, R., Volinsky, C.: Matrix Factorization Techniques for Recommendersystems. IEEE Computer 42(8), 30–37 (2009)
12. Marshall, I., Roadknight, C.: Provision of Quality of Service for Active Services. Computer Networks 36(1), 75–85 (2001)
13. Xu, L., Shi, K., Goto, Y., Cheng, J.: ISEC: An Information Security Engineering Cloud. In: Proceedings of the 3rd IEEE International Conference on Software Engineering and Service Science (ICSESS 2012), pp. 750–753. IEEE Press (2012)
14. Xu, L., Wang, B., Goto, Y., Cheng, J.: Providing Users with Suitable Services of Information Security Engineering Cloud Based on ISO/IEC 15408. In: Proceedings of the 4th IEEE International Conference on Software Engineering and Service Science (ICSESS 2013), pp. 321–325. IEEE Press (2013)
15. Zhang, N., Iqbal, A., Goto, Y., Cheng, J.: An Analysis of Software Supportable Tasks related with ISO/IEC 15408. In: Proceedings of 9th International Conference on Computational Intelligence and Security (CIS 2013), pp. 601–606. IEEE Computer Society Press (2013)

Identify the Online Product Comments with Suspicious Chinese Content

Ping Han Lei[1], Pingyu Hsu[1], and Ming Shien Cheng[1,2,*]

[1,2] Department of Business Administration,
National Central University, Taiwan
pyhsu@mgt.ncu.edu.tw
[2] Department of Industry Technology and Management,
Ming Chi University of Technology, Taiwan
mscheng@mail.mcut.edu.tw

Abstract. Recently, people who need more information about the goods they planned to purchase will look for the online product reviews before the purchasing. This is how "Electronic Word-of-mouth" (eWOM) influences or even changes the purchasing decision. The purpose of this research is to identify the worst kind of the online product reviews: rumors. Rumors could cause serious damage to company's goodwill and the sale of the product. In this study, we developed a new method that combined the research of rumors and the text mining techniques. Breaking the content of online product review into two components, and then use the "Keyword matching" technique to evaluate whether it is a rumor article. The result of this method shows that it could precisely identify those rumor articles from bunch of online product reviews. We could use it as a filter when we search for product information and make a better and more suitable buying decision. Based on the models developed in this study, the results show that the articles with more important attribute vocabulary and fuzzy vocabulary and fewer words are more likely to contain rumors. The results also show that rumor articles and articles containing normal responses to questions can be effectively separated. The collected training set results show: precision=71.43%, recall=73.5%, F-measure=72.45%; the testing set results show: precision=80%, recall=73.73%, F-measure=76.19%.

Keywords: Online word-of-mouth, Text Mining, Rumor, Internet Rumors, Vague Vocabulary, Online Product Review.

1 Introduction

In recent years, word-of-mouth marketing through the Internet and releases of unpacking articles by well-known bloggers, and sharing of product experiences have become increasingly popular. Consumers have also become more accustomed to collecting and viewing related articles on the Internet as references when choosing

---

* Corresponding author.

M. Chau et al. (Eds.): PAISI 2014, LNCS 8440, pp. 49–64, 2014.

products. In particular, when buying products with greater complexity and involvement, such as computer components, laptops, smartphones, tablet computers, and other technological products, they prefer to refer experts' appraisal articles and feedbacks from other consumers. However, many of the articles are written by someone hired by firms, thus the authenticity of the experiences and the objectivity of the insights are suspicious. Such viral marketing has advantages of low costs and wide spread, and they are more convincing than general marketing advertising. Hence, this type of marketing practice has been well received by firms or marketing companies. From simply promoting one's own product characteristics, articles that attack products of other competitors have gradually emerged. Writers as users of other competing products post negative experiences, make vigorous attacks, and even fabricate product defects, attempting to affect other consumers' purchase decisions by spreading rumors. Such rumors have flooded the network forums, which have affected consumers' purchase behaviors and purchase intentions. The good intentions behind experience sharing have become abused and twisted.

The purpose of this study is to prevent abuses of Internet viral marketing techniques that lead to the destruction of online forum impartiality and credibility, which in turn deceive consumers and provide erroneous information. Thus, text mining in information mining was adopted to establish a detecting model. The rumor and slander articles screened from the forum article using the established model shall serve as a reference for online forum in order to more efficiently manage. In this study, the Android smartphone discussion area of the well-known network forum Mobile01 (www.mobile01.com.tw) was adopted as the object, and two billboard articles on HTC and Samsung were targeted for analysis. At the end of 2012, the network forum users commented that the HTC discussion boards were filled with Internet writers hired by competitors to continue to post slander articles to attack HTC products, with the intent to create the negative image of high brand failure rate and low product quality. In April 2013, the user but with different names posted offensive articles against HTC and articles that promoted the advantages of Samsung phones. The said user both played the role of questioner and answerer and presented the shortcomings of HTC phones and recommended Samsung phones. He was caught by other forum users to have been asking and answering questions using the same account with different identities. In May 2013, a similar event of questions and answers by the same person took place again. Obviously, similar tactics on Samsung phone and HTC phone discussion boards were often appeared. In this study, based on the characteristics of rumors and the text mining techniques, the rumor composition was explored to establish a model for screening rumors and put forth a new type of classification method in order to analyze articles that are more likely to be fraud and articles that are suspicious.

This paper is organized as follow: Section 2 presents related work. Section 3 describes the methodology and model of our research. Section 4 evaluated our research result. Section 5 draws some conclusion and addresses the future work.

2   Literature Review

2.1   Rumors

Kapferer in 1992 [20] compiled past researches and redefined rumors: "Rumors are messages without official confirmation or that had been officially refuted and appeared and circulate within the community. The term "official" does not specifically refer to government organizations; it may refer to parties concerned or vendors. A rumor has at least three characteristics: (1) The rumor is relevant to the current facts; (2) the purpose of the rumor is for people to believe the contents, thus they present as much realistic scenario as possible; (3) the rumor emphasizes authenticity. Therefore, the contents may not be entirely wrong, they are just mostly unconfirmed messages. In general, rumors refer to unfounded and made-up stories. Scholars were inclined to define rumors as unproven messages. Shibutani in 1996 [30] mentioned that "rumors" start from important and unspecified events. Message receivers are likely to accept or believe a rumor due to the great importance of an event and their failure to obtain adequate knowledge or information to prove. Allport and Postman [10] [11] as the founders of rumor research have developed a conceptual equation consistent with the argument of Shibutani:

$$\text{Rumors} = \text{Importance of events} \times \text{Ambiguity of events}$$

Kapferer in 1992 [20] pointed out that according to the equation, if the importance of an event is zero, or if the event itself is not vague, rumors cannot be generated. Fisher in 1998 [16] believes that rumors are the circulation of information for a certain purpose, a kind of collective behavior among the masses.

In academia, the term "rumor" has been defined as a neutral concept, not entirely negative. "Rumors are generally intuitively associated with negative implications and are deemed as erroneous information. Most of Internet rumors are indeed erroneous messages, but this is not absolute, some rumors may eventually prove to be true, while real messages are spread similar to the way rumors are circulated" [1]. Rumors have a wide spectrum, DiFonzo and Bordia in 1994 [14] pointed out in their study that the impact of rumors includes tangible and intangible damage. Tangible damage includes a decline in product sales, which is the most direct problem; intangible damage refers to a company's reputation or customer trust.

Over the past, rumor-related researches mainly targeted the forms of rumors spread through word-of-mouth. The range of spread was limited to specific communities or situations. With the advancement and prevalence of the Internet, people's traditional communication channels and message-spread modes have undergone major changes. There is no barrier for posting comments on the Internet, the generation and dissemination of rumors have become more convenient than ever, and the Internet has become an important channel for spreading rumors. In recent years, social media like Facebook and Twitter have gradually become pipelines for rumors. Pasted and shared information are forwarded through convenient button before the information is proven true. Through the interpersonal networks, friendly reminders, and alarmist rumors, outdated and erroneous information continue to be passed to all corners of the world. Before a rumor can be refuted or confirmed, it may have been passed around for months or even years. Internet rumors may have negative effects on product sales, share prices, and corporate image [15]. According to Wang and Lai's research in 2001 [2] proposed that rumors about product defects cause significant harm to enterprises. Regardless of consumers' believing of rumor, their attitude toward the product may change due to rumor, thus affecting their consumption decisions.

2.2   Word-of-Mouth Marketing

Broadly speaking, word-of-mouth (WOM) marketing is evaluations from consumers distributed through any pipelines; it is an informal marketing approach that contains both positive and negative information. Blackwell in 2006 [13] defined WOM as the sharing and dissemination of product ideas, evaluation, and personal experience under the premise that consumers are without commercial intent. Bickart in 2001 [12] believed that compared to traditional marketing tools, WOM is more trustworthy and relevant, because WOM is easy to identify with, which can effectively reduce the marketing practices that result in customers' sense of resistance. Sernovitz in 2009 [30] also pointed out that WOM marketing is more likely to achieve publicity results compared to directly doing traditional marketing publicity on consumers.

Compared to traditional WOM, electronic word of mouth (eWOM) is much faster, which is mainly attributed to the more diverse forms of Internet communication. Graham and Havlena in 2007 [11] mentioned that the Internet features storages of records, coupled with multimedia applications, which dramatically improves the effect and visibility of eWOM. Litvin in 2008 [23] pointed out that when consumers make purchase decisions, the important referential information includes the effects of interpersonal relationships and eWOM. Riegner in 2007 [27] pointed out that the effect of eWOM is more obvious when making electronic product purchase decisions, likely due to the expensive prices, high complexity, and deep product involvement. Prior to making a purchase, the consumer will first collect adequate related information and refer to the opinions of other consumers. According to the finding of a survey in American, about 61% of the consumers first refer to comments on the Internet, blog, or the experiences and opinions of other consumers before buying a product [32]. Almost the same survey in Taiwan released by Insight Explorer in 2009 [9], as high as 63.1% of the Internet users agree that they highly valued eWOM before making a purchase and that they will first collect related information online before making a consumption decision.

2.3   Text Mining for Suspicious Content

According to the Humphery's research [28] that manual manipulation and fraud public report contents use more encouraging words, images, etc., with fewer changes in diction. Frauds are focused on the display of a reliable image to reduce descriptions of the actual situations. This study attempted to distinguish fraud financial reports based on content and grammatical structure using eight variables as the indicators of test reports: text influence, textual complexity, the degree of wording difference, the level of emotional expression, the level of indirect expression, syntactical function, sentence quantity, the quantity of special characters, and the ratio of uncertainty words. Among them, uncertainty terminology uses the ratio of quantity of modal verbs vs. the total number of verbs in messages as representative variables. The main function of modal verbs is to act as verb auxiliaries, such as possibility, necessity, etc. In the study, the English words include would, should, could, etc. In a sentence, the presence of this type of modal verbs will reduce the credibility of the sentence, which means the speaker or author is not confident about the content and cannot make a commitment.

Humphery's research showed that the Bayesian classification and C4.5 classification tree method work the best, while the use ratio variable of uncertainty words more related to the study shows no significant distinguishing effects. The reason is that senior managers that write reports have received training in accounting, so they know how to hide unfavorable information within the norms, without having to use uncertainty words to avoid it. The Internet rumors in this study are not prescriptive, and the message content is without any restrictions. Unlike the conditions of financial reports, writers have not received professional training, and the article content avoidance technique used is relatively simple.

Oberreuter and Velaïsquez in 2012 [17] studied text mining applied in detecting thesis plagiarism and finding writing style variations. Unlike traditional cross-checking methods, their study analyzed documents by text mining, and the writing style model was established based on the linguistic features, and paragraphs with different writing styles were found from the entire article, which could have been written by someone else. This method requires no crosschecking procedures, and the language used in the article is irrelevant. With the self-based information of the articles, the paragraphs with excessive differences were deemed outliers and were picked out. The results show that the algorithm developed in this study is more effective than commonly used methods in the past. Moreover, due to the simplified process, it is much easier to operate. However, the results with relatively low precision show that this method may not be perfect enough although it is currently the most effective model in the field. Specially, this method applied on non-English articles, the starting point and stop words of text must first be pointed out. Word segmentation must be done for articles in Chinese.

The method proposed in this study is intended to identify the outlier of the article content. Unlike the method in our study, the classification objects are those that were distinguished in the same writing style. The purpose of our study is to classify specific articles from corpuses by different authors. Due to the complexity and diversity of the writing styles, the traditional crosschecking method was adopted as the classification method.

2.4   Rumors with Vague Words

Wang and Luo in 2002 [1] conducted an analysis on the Internet rumors in the country through the content analysis method. It was pointed out that the content form of the Internet rumors conformed to the two elements stated by Kapferer [20]: the importance of events and fuzziness of events. As for the framework, the three basic components of rumors proposed by Koenig [21] were adopted for integrative analysis. The components include (1) the target; (2) the charge or allegation; (3) the source. The analysis results show that 46.3% of the Internet rumors were intended for specific objects such companies and groups (34.7%) or individuals (11.6%); 34.7% of the rumors only mentioned vague time words, such as "once" , "one day" and "last month"; 47.7% of the rumors were without specific time, targeting only the event or phenomenon that continues to occur. As for sources of rumors, 66.3% heard it from someone else and only 33.7% were the source of the rumor. As for the analysis of the charge or allegation of rumors, up to 31.6% intended to change message receivers' purchase decision which refuse products or services of specific company. Rumors of this type cause harm to companies or groups and bring negative effects to their image and reputation.

Lakoff in 1973 [22] defines "hedging" as a term that increases or decreases vagueness. After compiling the definitions of hedging by a few scholars, Xu in 2007 [3] mentioned about Kopple [31] who deems the reason to use vague vocabulary in articles is that "the speaker does not fully guarantee the propositional content of his own words." The usage of "seem," "perhaps," and "might" is to adjust the truthfulness of the whole proposition, rather than expressing the uncertainty of a single constituent in the sentence. Chen in 2008 [4] believed that today's hedging generally refers to the speaker's varied degrees of response for the proposition expressed, which is an important mechanism for mitigating the responsibility of what one says.

Perkins in 1983 [25] believed that modality originated from people's lack of knowledge and failure to assert the truthfulness of an event or proposition. They merely change their way of saying similar contents. Lyons in 1977 [24] defined modality as the proposition of conversational content and the description of the speakers' viewpoint or attitude, which can be explained by the concept of the "possibility" and "necessity" logic. Furthermore, when explaining epistemic modality, modality and hedging share very similar definitions as they both possess vagueness. Vague vocabulary can reduce conflicts in conversations.

3   Rumor Model

3.1   Methodology

In this study, text mining technology was combined with Chinese rumors research using the keyword comparison method as the main basis for the classification of Internet articles, and according to the conceptual formula proposed by Allport and

Postman [10]: a model was developed to classify the articles with the nature of false rumors. In this study, the event importance is defined by the attributes and functions of smartphones valued by consumers, because it is closely related to their experience. And consumers usually collect and assess information before making a purchase decision. Event vagueness is represented by modal adverbs in ambiguity or possibility vocabulary often used in Chinese writing. The vaguer the language that appears in the sentences, the more it represents great vagueness, and implied the author cannot guarantee the message at all.

The process of our model described as follows: firstly, retrieve product review articles from the Internet forum as input data, then established the rumor prediction model. This rumor model can be the preliminary step of filtering out rumor articles. The study process is illustrated below:

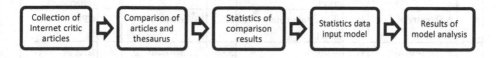

Fig. 1. The process of building model

3.2　Keywords Retrieval

In the keyword retrieval method, we choose the thesaurus comparison method. Totally two thesauruses about smartphone were established: important attributes thesaurus and vague language thesaurus. The process to establish an important attributes thesaurus is divided into two steps. Firstly, according to the smartphone-related research [8] and smartphone professional appraisal websites [33], the important attributes of smartphone include six categories: exterior design, operational function, multimedia, call and quality, endurance, memory capacity, etc. Then three experts were assembled to give advice by content analysis method which is commonly applied in qualitative research. Based on the previously listed smartphone's important attributes which were extracted from this study's collected articles, we would create an important attributes thesaurus. The three experts were selected from the members of forum which is the discussion boards of Mobile01 Forum of smartphones, and they have been concerned about the Internet rumors issues, especially possessed in-depth understanding of the vocabulary frequently appeared in smartphones and Internet Forum.

Three experts chose words for smartphone attributes and specifications which consumers concerned when making purchasing decisions, included the terminology and special description from collected articles. They joint decisions to add or delete words by voting system that at least two experts agreed should be incorporated into the thesaurus. The final outcome is the important attributes lexicon containing 87 important attributes of smartphone as follows:

Table 1. Important attribute thesaurus

No.	Item	No.	Item	No.	Item	No.	Item
1	喇叭(speaker)	23	色差(color difference)	45	邊框(frame)	67	對焦(focus)
2	NFC	24	黑屏(black screen)	46	充電蓋(charge cover)	68	漏光(light leak)
3	螢幕(screen)	25	液晶(liquid crystal)	47	背蓋(back cover)	69	電池(battery)
4	亮點(bright spot)	26	AMOLED	48	外殼(shell)	70	充電(charging)
5	透光(light)	27	LCD	49	邊條(edge strips)	71	機板(board)
6	黑點(black spots)	28	圖形鎖(graphic lock)	50	材質(material)	72	主機板(motherboard)
7	黃斑(macular)	29	外觀(outward appear-ance)	51	背殼(back-shell)	73	處理器(processor)
8	入塵(entry of dust)	30	鋁合金(aluminum alloy)	52	鏡頭(lens)	73	CPU
9	偏光(polarized)	31	縫隙(slit)	53	閃光燈(flash lamp)	74	記憶體(memory)
10	烙印(imprint)	32	縫(stitch)	54	輔助燈(auxiliary lamp)	76	RAM
11	觸控(touch)	33	返回鍵(return key)	55	藍牙(blue-tooth)	77	訊號(signal)
12	USB	34	按鍵(button)	56	藍芽Blue-tooth	78	收訊(reception)
13	記憶卡(memory cards)	35	鍵盤(keyboard)	57	系統(system)	79	聽筒(earpiece)
14	SD	36	電源鈕(power button)	58	更新(update)	80	APP
15	WIFI	37	電源鍵(power button)	59	開關(switch)	81	OS
16	SIM card	38	麥克風(microphone)	60	通話(conversation)	82	UI
17	充電孔(charging hole)	39	韌體(firmware)	61	撥電話(dial phone)	83	Sense
18	傳輸孔(transmission hole)	40	軟體(software)	62	播放器(players)	84	玻璃(glass)
19	耳機孔(headphone jack)	41	介面(interface)	63	3G	85	LED
20	相機(camera)	42	頓頓的(stacked up)	64	GPS	86	LTE
21	HOME	43	卡卡的(pileup)	65	Blue-tooth	87	4G
22	HOME鍵(home button)	44	卡頓(pileup and stacked up)	66	瀏覽器(browser)		

Source: This Study

The creation of the ambiguity attributes corpus is based on integration of Chinese scholars' summarized modal particles and possibility adverbs. In the Chinese language related study, ambiguity phrase mainly existed in modal adverbs category. Peng in 2009 [5] integrated various scholars' studies of English modal adverbs and

Chinese modal adverbs, and proposed that Chinese possible tone, speculation tone, and judging tone all express the assessment of proposition authenticity, which meant that there were words of highly reliable such as "一定(certain)", "必然(certainly)", "必定(will)" and so on, and words of lowly reliable such as "也許(may)", "大概(probably)", "或許(might)" and so on. Xu in 2007 [6] cited seven modal adverbs in the speculation class: "大概(probably)", "似乎(seems)", "或許(may)", "也許(might)", "應該(should)", "好像(like)", and "可能(possible)", which were consistent with possibility adverbs: expressing the speaker's attitude of uncertainty or to avoid commitment. The study found that the mixed use of speculation class modal adverbs will weaken the speaker's commitment to the speech. Chen's study in 2009 [7] integrated Western and Chinese relevant discussion and classification methods from the modality's point of view, and identified the modal category corresponding to "probability" in Chinese. Furthermore, he selected the seven possibility adverbs: "應該(should)", "也許(may)", "或許(perhaps)", "好像(like)", "似乎(seems)", "可能(possible)" and "大概(probably)" by summarizing Chinese scholars' study and modern Chinese corpus frequency. Finally, the ambiguity attributes corpus was as follows:

Table 2. Vague vocabulary thesaurus

No.	Vocabulary
1	大概(Probably)
2	似乎(Seem)
3	或許(Perhaps)
4	也許(May)
5	應該(Should)
6	好像(Like)
7	可能(Possible)

Source: [6][7]

Data mining software was used to build the model, and two thesauruses were used in the comparison with the study's collected articles. When the article vocabulary matched the thesaurus words, the system automatically marked and presented the results. For example: "My HTC phone screen may be faulty." The thesaurus compared "screen" and "may", and marked the times appeared in article to be 1 each word. The total number of words of each article matching the important attributes thesaurus and vague vocabulary thesaurus were summarized. In order to avoid the final model resulting in zero and thus being unable to properly assess the articles, the initial value of the number of matched vocabulary was set to 1 to facilitate subsequent calculations and evaluation.

3.3   Model

In the modeling process, we referred TF-IDF evaluation concept which was commonly applied in data mining to assess the importance of the vocabulary in the document.

The TF value of the vocabulary is how frequently the words appeared in the document. The higher value represents the higher importance of the vocabulary. IDF value is the ratio of the total amount of documents to the total number of times the vocabulary appears in all documents. The higher IDF values represent the lower importance of the vocabulary, which may be the vocabulary used in common documents.

The study's established thesaurus through the integration of experts' opinions screened vocabulary closer and more important to the study, but does not contain general terms used in common documents. Only the frequency was taken as the indicator to evaluate the importance of the vocabulary. The total number of important attributes terms has a larger difference between the articles, so its logarithm is taken to make adjustments, making the effect of this study model more remarkable. Finally, a model was developed where the higher the evaluation values of the rumors, the greater the likelihood of the articles being rumors. The model formula is as follows:

$$rumor\ evalution\ value$$
$$= \frac{(number\ of\ ambiguity\ attribute\ phrase + 1)}{total\ number\ of\ words\ in\ article}$$
$$\times log(number\ of\ importance\ attribute\ phrase + 1)$$

In this study, the keywords were compared with the content of the article to calculate its precision, recall, and F-measure as a basis to assess the quality of model effect. Precision: correct classification rate of rumors in the articles.

$$(Precision) = \frac{total\ number\ of\ correct\ classification\ as\ rumor\ articles}{total\ number\ of\ classified\ rumors\ artilces}$$
$$\times 100\%$$

Recall: the proportion of the classification results as rumors in total rumor articles.

$$(Recall) = \frac{total\ number\ of\ correct\ classification\ as\ rumor\ article}{total\ number\ of\ rumor\ articles} \times 100\%$$

F-measure: The average score concept is that the higher the precision and recall, the higher the average score, which means the better the classification results will be. Generally, the highest point was taken as the standard for assessing the effectiveness of model, and the corresponding rumor assessment value as the threshold of the present study.

$$F - measure = \frac{2 \times Precision \times Recall}{Precision + Recall} \times 100\%$$

4   Experiments and the Evaluation

4.1   Data Collection

The data source is from the largest Internet forum of electronic consumer products in Taiwan–Mobile01 smartphone discussion boards. In particular, the articles extracted from the HTC mobile phone discussion boards and Samsung mobile phone discussion boards, because these two brands are the most mainstream brands in Taiwan's Android market. Discussion boards are rich in articles and have numerous participants in discussions. The articles mainly discussed new machine evaluation or mobile failure problems, etc. Such environment is suitable for eWOM and viral marketing. Moreover, manufacturers often spread unfavorable rumors to competitors, so the articles about experience and fault conditions of these two discussion boards were taken as the main objects of study.

4.1.1   Filter Criteria of Manual Collection Articles

Data collection method was the manual collection of articles in the discussion board about phone failure, the questions of usage and complaints, which may cause the product's negative publicity effect, including the articles that normally reflect the problem and are more likely to be rumor articles. This study provides experts reference information on whether it is an article of normal reflection of the problem as follows: (1) whether it complies with the posting provisions of forum: reacting to product failure problems or defects must be attached by photo evidence, and maintenance treatment must be attached by maintenance documents. (2) The same problem also occurred in other articles, and replied in support of the argument or reaction. (3) The article was locked, warned, and even pruned by the forum administrator would be included in the rumors. (4) Refer the sender's account information, such as past posting history, time to establish the account, and the proportion of account number posting on each forum.

4.1.2   Classification of Articles with Experts' Opinions

This study collected 110 Internet articles and reference information, and classified by three experts through a voting system to determine whether the article would be a rumor. It is classified as a rumor article as long as two out of three votes agreed. In the circumstances of only one expert deeming it a rumor, the three experts have to carry out a discussion and re-vote, to achieve a consensus, more than two must agree the classification results. Classified in this way and reducing the degree of disagreement, the three experts' opinions integrated the final classification result: a total of 56 articles are rumors, and a total of 54 articles are of normal reaction to the problems.

4.2   Experiment

Before conducting the thesaurus comparison, the classified rumors and normal articles were first randomly staggered, and then 30% of random numbers were extracted as the testing data: 36 articles, and 70% as the training data: 74 articles. This study used SPSS Clementine for thesaurus comparison, and results entered in Microsoft Excel. The statistical results of each article contained three fields: the number of important attributes vocabulary, the number of vague vocabulary, and the total number of words of individual article. These three field values were brought into this study to develop the model formula:

rumor evalution value

$$= \frac{(number\ of\ ambiguity\ attribute\ phrase + 1)}{total\ number\ of\ words\ in\ article}$$

$$\times \log(number\ of\ importance\ attribute\ phrase + 1)$$

In order to calculate the individual articles of the assessed value of the rumors, training data calculations are first sorted in descending order, to facilitate the calculation of precision, recall, and F-measure and set a threshold. The greater the assessed value, the greater the possibility of those articles being rumors; otherwise, the greater the likelihood of being a normal post. The results obtained in this study show an accuracy of a descending curve, while the recall shows an ascending curve. If the threshold is set at a higher place, the classification result obtained will be high precision but relatively low recall. If the threshold is set lower, the result will be high recall but low precision. If the gap is too large, it will not be suitable for use, so in consideration of both precision and recall, F-measure is taken as the evaluation indicator.

This study's developed model was used to conduct classification of rumors, in which 74 training data were computed with the model. The results showed below:

Fig. 2. The experiment result of training data. (Source: This study)

Figure 2 indicates that the precision is a descending curve, and recall is an ascending curve. At the highest point of the F-measure curve, the precision is very close to the recall. If it is taken as the standard for classification, there will be no situation where precision and recall is poorer. Therefore, this study took the highest point 72.45% of F-measure as the best classification results with a precision of 71.43%, and recall of 73.5%. Rumors assessed value of this point was 0.003679833 and taken as the threshold. In testing data, the one with rumors assessed value greater than the threshold value was classified as a rumor article.

Table 3. The evaluation result of trainning data

Index	Precision	Recall	F-measure
Data	71.43%	73.5%	72.45%

The results of model operations of 36 testing data in this study are as shown below:

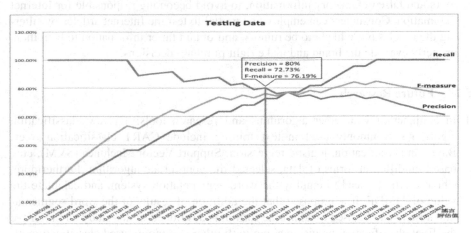

Fig. 3. The experiment result of testing data (Source: This study)

The derived threshold of 0.003679833 from training data would be the standard. The minimum rumors assessed value of the testing data above the threshold was 0.003861713; precision was 80%; recall was 73.73%; F-measure was 76.19%. To make this threshold value as the classification criteria of the testing data, its effectiveness of precision and recall was good, and the difference was not significant.

This study focuses on screening rumor articles from the numerous articles, paying more attention to the effectiveness of precision, and the results of the precision of testing data reach 80% indicates that the model can effectively sort out rumor articles. Such experimental results conformed to the anticipation of this study.

Table 4. The evaluation result of testing data

Index	Precision	Recall	F-measure
Data	80%	73.73%	76.19%

5   Conclusion and Future Research

5.1   Conclusion

1. Internet rumors aimed at product or manufacturers often appeared in the Internet. Through a far-reaching network and timeless characteristics, rumors will bring great negative impact to product image or manufacturers reputation. How to effectively curb rumors is quite an important issue for the enterprise. This study targeted

Internet rumors of consumer products to conduct analysis on content. A model developed in this study found that the more vague vocabulary and important attributes, and less total number of words in the article, the higher the possibility of rumor. The findings also show that the model can effectively separate the rumors from normal article.

2. Internet forum managers can apply this model to filter forum article, then clear rumors and false defamatory information, to avoid becoming responsible for Internet defamation. Consumers can employ this model to test the Internet articles by filtering the articles more likely to be rumors, and obtain fairer information to help them properly evaluate the brand and make right purchase decisions.

5.2   Future Research

1. In conjunction with other algorithms and research tools: A few classification algorithms commonly used in text mining include CART classification trees, Bayesian classification, logistic regression, Support Vector Machines (SVM), etc., based on different forms of data, selected the appropriate algorithm. Furthermore, Chinese articles need to employ the word segmentation system, and calculate the value of TF-IDF. It is proposed that the addition of results of the word segmentation system can be used to adjust the formulas developed in this study and improve the final classification precision along with other commonly used classification algorithms.

2. Increase of different product type research: This study only focused on specific brands of smartphones, and whether there are differences in the findings on different brands of smart phones? In terms of other consumer electronics products, such as digital cameras, tablet PCs, or rely on eWOM such as cars, would be the different results? These objects can be explored in future researches. It is proposed that comparison of high and low involvement of other consumer electronic products could be the research scope.

3. Expand the breadth and depth of research data: In this study, the article data setting range are articles published within one year from April 2011 and belong to the Samsung smartphone and HTC smartphone boards. In terms of the breadth, the future research can be extended to boards of other smartphone brands or even include articles of all smartphone discussion boards. In terms of depth, the focus in the future can be on all of the discussion articles of two discussion boards, the timeline of the research was extended for in-depth study of the discussion board.

References

1. Allport, G.W., Postman, L.: An Analysis of Rumor. Public Opinion Quarterly 10, 501–517 (1947)
2. Allport, G.W., Postman, L.: The Psychology of Rumor. Henry Holt, New York (1947)
3. Bickart, B.A.R.M.S.: Internet Forums as Influential Sources of Consumer Information. Journal of Interactive Marketing 15(3), 31–40 (2001)
4. Blackwell, R.D., Miniard, P.W., Engel, J.F.: Consumer behavior. Aufl.Mason 10 (2006)

5. Chen, I.T.: The usage of hedging words on Chinese conversation (in Chinese), Master thesis: Graduated School of Linguistics. National Taiwan University (2008) (unpublished)
6. Chen, Y.C.: The research of possibility ranking of modern Chinese adverbs (in Chinese), Master thesis: Graduate School of Chinese Language Teaching. National Taiwan Normal University (2009) (unpublished)
7. DIfonzo, N., Bordia, P., Rosnow, R.L.: Reining in rumors. Organizational Dynamics 23 (1994)
8. Difonzo, N., Bordia, P.: How top PR professionals handle hearsay: Corporate rumors, their effects, and strategies to manage them. Public Relations Review 26(2), 173–190 (2000)
9. Fisher, D.R.: Rumoring Theory and Internet-A framework for Analyzing the Grass Roots. Social Science Computer Review 16(2), 158–168 (1998)
10. Graham, J., Havlena, W.: Finding the missing link: Advertising's impact on word of mouth, web searches, and site visits. Journal of Advertising Research 47(4), 427–435 (2007)
11. Hoffman, D.L., Novak, T.P.: Marketing in Hypermedia Computer-mediated Environments Conceptual Foundations. Journal of Marketing 60(3), 50–68 (1996)
12. Humpherys, S.L.: Identification of fraudulent financial statements using linguistic credibility analysis. Decision Support Systems 50 (2011)
13. Hung, P.H.: Key factors of decision-making of buying smartphones, Master thesis: Department of Industrial Management. National Taiwan University of Technology (2011) (in Chinese)
14. Hsu, C.N.: Modal discourse of modern Chinese. Kunlun Press, Beijing (2007) (in Chinese)
15. Hsu, H.L.: The research of stance markers of academic discourse. Hefei University of Technology, Hefei (2007) (in Chinese)
16. InsightXplore Market Research Consultant, Internet word of mouth marketing ram consumption decisions about eighty percent of users, 10 (2009) (in Chinese), http://www.insightxplorer.com/news/news_10_30_09.html
17. Kapferer, J.N.: Rumors- Uses, Interpretations, and Images. Transaction Publishers, New Brunswick (1992); Zheng, R.-l., Bien, Q. (trans.) Laurel Book Store Published, Taipei (1992)
18. Koenig, F.: Rumor in the Marketplace. Auburn House, Dover (1985)
19. Kopple, W.J.V.: Some exploratory discourse on metadiscourse. College Composition and Communication 36, 82–93 (1985)
20. Lakoff, G.: Hedges: A study of meaning criteria and the logic of fuzzy concepts. Journal of Philosophical Logic 2(4), 458–508 (1973)
21. Litvin, S.W., Goldsmith, R.E., Pan, B.: Electronic word-of-mouth in hospitality and tourism management. Tourism Management 29(3), 458–468 (2008)
22. Lyons, J.: Semantics, vol. 1&2. Cambridge University Press, Cambridge (1977)
23. Oberreuter, G., Juan, D., Velaïsquez: Text mining applied to plagiarism detection: The use of words for detecting deviations in the writing style. Expert Systems with Applications 40 (2013)
24. Opinion Research Corporation, Online consumer reviews significantly impact consumer purchasing decisions (June 2008), http://www.opinionresearch.com/fileSave/Online_Feedback_PR_Final_6202008.pdf
25. Pendleton, S.C.: Rumor Research Revisited and Expanded. Language & Communication 18, 69–86 (1998)
26. Peng, Y.C.: The Analysis of leading code of Chinese modal adverb, Master thesis: Graduate School of Chinese Language Teaching. National Taiwan Normal University, Unpublished (2009) (in Chinese)

27. Perkins, M.R.: Modal expression in English. Ablex Publishing Corporation, Norwood (1983)
28. Riegner, C.: Word of mouth on the web: The impact of Web 2.0 on consumer purchase decisions. Journal of Advertising Research 47(4), 436–447 (2007)
29. Sernovitz, A.: Word of Mouth Marketing. Kaplan Publishing, New York (2009)
30. Shibutani, T.: Improvised News: A Sociological Study of Rumor. Bobbs Merrill, Indianapolis (1966)
31. Top Ten Reviews - Smartphones (Website),
    http://cell-phones.toptenreviews.com/smartphones/
32. Wang, C.C., Lo, S.K.: The research of exploring Internet rumors by content analysis. Journal of Information, Technology and Society (2002) (in Chinese)
33. Wang, C.C., Lai, M.C.: Consumer's attitude of spreading defective products rumors on electronic bulletin board. Advertising Research 16, 31–52 (2001) (in Chinese)

Automatically Explore Inter-Discipline Technology from Chinese Patent Documents

Ming Shien Cheng[1,2,*] and Pingyu Hsu[1]

[1,2] Department of Business Administration, National Central University, Taiwan
pyhsu@mgt.ncu.edu.tw
[2] Department of Industry Technology and Management,
Ming Chi University of Technology, Taiwan
mscheng@mail.mcut.edu.tw

Abstract. In knowledge-based economy, technologic capability would be the core assets of enterprises, and technology management is considered one of the most important management activities. The foremost work of technology managers should be exploring technology trends to support technology planning. However, exploring technology trend is no longer concentrated in one disciplines, much innovative research results from integrating technologies developed in different disciplines. An automatic tool to help explore inter-discipline technology should be very useful for technology managers. Recently, some researches which through the patent documents to explore technology trend mostly identify the trend by patent map, described the technology trend each discipline separately, and can't explore inter-discipline trend to help technology manager broader vision. Moreover, most of discussing technology trend studies, even discovering emerging technology—the most popular topic, adopted the published patent documents written by English and based on USPTO for patent analysis. Seldom of patent analysis research collected patent documents written by Chinese—the most population language. Our research attempts to remedy the shortfall of aforementioned studies, and propose an integrated technique. The research utilizes common themes development methodology to identify multiple discipline and the methods work with patents written in Chinese. Our developed method does not try to predict future development, but try to present current technology development as it is in a succinct way to help technology managers make the judgment. Our system would provide inter-discipline technology analysis which across transmission technology and transmission of digital information technology, and the empirical experiment found twenty three inter-discipline technological subjects. Furthermore, extended experimental scope—added one discipline (image communication technology), our experiment found thirteen inter-discipline technological subjects.

Keywords: Inter-Discipline Technology, Text Mining, Cross Collection Mixture Model, Chinese Patent Document.

---

* Corresponding author.

M. Chau et al. (Eds.): PAISI 2014, LNCS 8440, pp. 65–77, 2014.

1   Introduction

In knowledge-based economy, technologic capability would be the core assets of enterprises, and technology management is considered one of the most important management activities, for a remarkable high technology company might spend over 15% of revenue on Research and Development (R&D) [21]. In 2012, over US$450 billion was spent on R&D projects [22][23]. With rapid technology development, how to master the technology planning and avoid wasted investments had become the most important issue for technology managers. The foremost work of technology managers should be exploring technology trends to support technology planning. With the technology trend, technology managers could decide which projects or products should obtain priority supports, and resolve which technology improvement would be appropriate for company to invest.

However, the adoption of technology is no longer concentrated in single discipline, much innovative research results from integrating technologies developed in different disciplines. As a result, technology managers must have macro and cross discipline perspectives when analyzing technology trends. "With the explosion in availability of information, scientists and technologists find it increasingly difficult to remain aware of advances within their own disciplines, much less in other seemingly unrelated ones."[6]. "Scholars have become increasingly aware of the need to link disciplinary fields to more fully answer critical question, or to facilitate application of knowledge in a specific area". [1] Technology manager requires broader technical knowledge than just the target discipline alone. [7] As a result, automatic tools to help explore inter-discipline technology should be very helpful to technology managers [6]. A tool that shows such a technology development trend should be able to save the time of browsing through tones of documents to identify the chances. The last but not the least, the tool should also be able to process Chinese documents since the market there is fast growing.

Recently, some research described technology trend by roadmaps [9] or lattice [8] with time variation to develop R&D strategy or monitor technological changes, but the technology trend describing each discipline separately, they can't describe inter-discipline trend to help technology manager broader vision. Most of discussing technology trend studies, even discovering emerging technology [2][4][5][9][17]—the most popular topic, adopted the published patent documents written by English and based on USPTO for patent analysis. Seldom of patent analysis research collected patent documents written by Chinese—the most population language. Our research attempts to remedy the shortfall of aforementioned studies, and propose an integrated technique. The research utilizes common themes development methodology [20] to identify technologies cross multiple disciplines and the methods work with patents written in Chinese. The developed method does not try to predict future development and identify research vacuums but try to present current technology development as it is in a succinct way to help technology managers make the judgment.

In this study, four categories patent documents (IPC classification code: H04N, H04B, H04L, and G06Q) of the WEBPAT Taiwan are collected. Among them, one is transmission, one is transmission of digital information, one is image communication the other is data processing system for administration. All patent documents were divided into several compared sets according to the published years (2003-2008). Then, we preprocess all patent documents by Chinese word segmentation system, and organize Chinese word with theme model. Finally, our system would provide inter-discipline technology analysis which across transmission technology and transmission of digital information technology, and the empirical experiment found twenty three inter-discipline technological subjects. Furthermore, extended experimental scope— added one discipline (image communication technology), our experiment found thirteen inter-discipline technological subjects.

The rest of the article is organized as following: Section 2 covers a review of researches on patent analysis and inter-discipline research methodology. Section 3 described research model design. Section 4 details the research data acquisition process and uses graphics to show the results. Section 5 summarizes the main contribution of this study and recommendations for future research.

2   Literature Review

Patent documents are an ample source of technical and commercial knowledge and, thus, patent analysis has long been considered a useful vehicle for R&D management and techno-economic analysis [19]. Automatic tools for assisting patent engineers or decision maker in patent analysis are in great demand [18]. According to many surveys of authorities, patents cover more than 90% of the latest technical information in the world, and 80% of the patent information is not published in any other form [13]. Patent analysis utilizes visualization interface: network[19][5], lattice[8], roadmaps [9], etc…, collectedly called patents map[17][10][12] to support technology manager strategic planning: discovering competitive intelligence[16], new product designing[12], uncovering new technology opportunity[10] or emerging technology [2][4], detecting technological capabilities[9], and exploring technology trend [8][13][19]. Traditional studies of patent analysis focus on single discipline, seldom of them cross analyze inter-discipline technology trend. We think inter-discipline technology trend analysis could broader technology managers' vision to make better decision.

Park et al. in 2012 [13] classified into two approaches of patent analysis tools, one is the bibliographic approach and the other is content-based approach. The bibliographic approach cannot identify detailed technological features and significant insights because it mainly relies on bibliographic information, which is considered to be superficial data. Content-based approach emphasizes technologically significant patterns, trends, and opportunities by extracting useful information such as abstracts, detailed description of invention and claims from patent text[18][8]. One representative content-based approach method is keyword-based analysis (KWA). Many researchers have developed keyword-based patent intelligence tools to identify trends in high-technology [19], to discover new technological opportunities from patent vacuums [9], to forecast new technological concepts [19], and to develop technology roadmaps [10]. Despite its simplicity and ease of use, KWA is limited in single discipline (collection) of documents. Our study commits to explore technology trend of inter-discipline (collection), so we adopt Cross-Collection Mixture Model (CCMM)

[20], which can discover the different common themes across all the collection and characterize what is in common among all the collections and what is unique to each collection. For exploitation of CCMM, Our study explores themes of inter-discipline and time varying, and display the technology trend by roadmap and cross-analysis.

3   System Design

3.1   Overall Research Process

Fig.1 depicts the overall research process, which consists of several stages. Firstly, some categories patent documents were collected from WEBPAT Taiwan database. Secondly, since patents are composed in Chinese language forms, we preprocess all documents by Chinese word segmentation system. Thirdly, we organize these words into three themes: background, common, and specific theme with CCMM and EM algorithm for extracting essential words to further analyze. Finally, our research provides inter-discipline technology analysis on common themes.

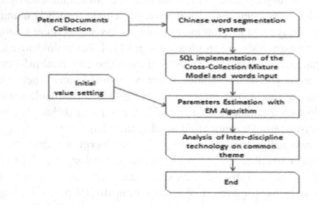

Fig. 1. Overall Research Process

3.2   Data Collection and Extraction

Patent documents in a technology field are collected based on various categories from WEBPAT. The patent documents need to be pre-process since they are semi-structured data, which are merely expressed in text format. For patent documents information, abstract would be most important part to briefly represent technology concept, so we collected summary of all patent documents collection.

3.3   Chinese Word Segmentation System Processing

Unlike English preprocessing steps of text mining, e.g. stemming, removing stop-words, etc., Chinese has word segmentation issues. The word is the most basic unit in preprocessing natural language, each "word" in English has its own meaning and a space to separate it from others, but as Chinese words have no spaces separating them, the Chinese word segmentation became the basic work for text mining. The current Chinese word segmentation systems have their own advantage; users can choose different systems on their needs. For instance, Academia Sinica's Institute Chinese word segmentation system provides online real-time segmentation function and has the new word recognition ability and selective function of the additional POS tag. In general articles, there are 3% to 5% of words are unknown, the abundance of the corpus in word segmentation system also determines the correctness of the results.

3.4   Processing Cross-Collection Mixture Model

In information retrieval and text mining, it is quite common to use a word distribution to model topics, or themes in text [3][11][14][15][20]. Following [20], we define different kinds of themes as follows:

Definition 1: Theme (θ): A theme in a text collection Ci is a probabilistic distribution of words characterizing a semantically coherent topic. Formally, a theme is represented with a unigram language model θ, i.e., a word distribution {p(w|θ)}w∈V s.t. Σw∈V p(w|θ)=1. High probability words of such a distribution often suggest what the theme is about. In this definition, we assume that the basic meaningful unit of text is a word, which is generated independently of each other. This definition can be generalized to adopt multi-word phrases as meaningful units. To model the temporal patterns of themes, we define the concept of three theme patterns as follows:

Definition 2: Background theme (θ_B): If a theme across all documents collections, we named it: Background theme. If a word appears a great number of times in a document as well as in all other documents, then that word is too common and is not representative. Researchers can collect the words that are too general and not representative in the background theme, Background theme replaced the manual processing of stop-words removal.

Definition 3: Common theme (θ_i): The common theme can be used to collect the continually existing and specific terms across documents collections of some interval time. We assume there are k latent common themes in all collections, and each is characterized by a multinomial word distribution (also called a unigram language model). A document is regarded as a sample of a mixture model with these theme models as components. We fit such a mixture model to the union of all the text collections we have, and the obtained component multinomial models can be used to analyze the common themes and differences among the collections. Formally, let C = $C_1 \cup C_2 \cup \cdots \cup C_m$, be m temporal collections of documents. Let θ_1, θ_2,..., θ_k be k theme unigram language models and θ_B be the background theme for all the collections. A document d is regarded as a sample of the following mixture model (based on word generation).

$$p_d(w) = \lambda_B p(w|\theta_B) + \Sigma_{j=1}^{m}[\pi_{d,j} p(w|\theta_j)] \tag{1}$$

where w is a word, $\pi_{d,j}$ is a document-specific mixing weight for the j-th aspect theme, and $\Sigma_{j=1}^{m} \pi_{d,j} = 1$. λ_B is the mixing weight of the background theme θ_B.

Definition 4: Specific theme: Collect words only appearing in a specific time interval. Our main idea for classifying themes for temporal text mining is to explicitly distinguish common theme clusters that characterize common information across documents collections of some interval time from special theme clusters that characterize collection-specific information. Thus we now consider k latent common themes as well as a potentially different set of k collection-specific themes for each collection (Figure 2). These component models directly correspond to all the information we are interested in discovering. The sampling distribution of a word in document d (from collection Ci) is now collection-specific. Specifically, it involves the background model (θ_B), k common theme models (θ_1, θ_2, ..., θ_k), and k$\times$i collection-specific theme models ($\theta_{1,i}$, $\theta_{2,i}$, ..., $\theta_{k,i}$), which are to capture the unique information about the k themes in collection C_i.

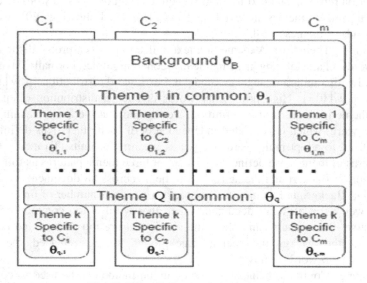

Fig. 2. Cross Collection Mixture Model

This study uses the Cross-Collection Mixture Model and conditional probability distribution to collect words. We express all word probability distributions as follows:

$$p_{d,c_j}(w) = \lambda_B p(w/\theta_B) + (1 - \lambda_B)(\sum_{\theta_i \in \Theta} \pi_{jdi} ((1 - \lambda_s)p(w/\theta_i) + \lambda_s p(w/\theta_{ij})) \quad (2)$$

The feature of the background θ_B parameter is that the word collected has a high frequency of occurrence in all documents. It means that the word is too common and can't represent the document, and the direction of patent research. The researchers can also set θ_B according to the experimental data to adjust the collected words of background. The determination of the common theme is to collect words with the overall representation in continuous time, so under the common theme's words are obvious and continuous occurrence in each document.

3.5   Parameters Estimation with EM Algorithm

This study used the Expectation Maximization (EM) Algorithm to calculate the relative maximum probability value of each word falling into each theme. This algorithm can help us repeat the calculation of the possible probability values of each sample in each theme and obtain the maximum possible values. The EM algorithm can identify the parameter maximum estimated value in the probability model, which must rely on hidden variables that cannot be observed. After the initialization parameter, it repeats the E step and M steps until convergence. E-step: To estimate the expectation value of unknown parameters, giving the current parameter estimates; M-step: to re-estimate the distribution parameters to make maximum likelihood of data, giving the expected estimates of unknown variables. The Expectation Maximization Algorithm is most commonly used in data mining clustering. This model also makes use of the theme to conduct clustering in order to make most similar documents fall into a similar theme.

In this study, based on the EM algorithm to express the words contained in all patent documents $C = \{d_1, d_2, ..., d_k\}$ belonging to the θ_B probability as follows:

$$P(w|\theta_B) = \frac{\sum_{j=1}^{m} \sum_{d \in C_j} c(w,d)}{\sum_{j=1}^{m} \sum_{d \in C_j} \sum_{w \in V} c(w,d)} \tag{3}$$

This probability formula represents the probability of the word w appearing in the sum of words in all possible document collections. The Maximum Likelihood Estimate was then adopted to find the parameter values.

$$P\left(Z_{d,C_j,w} = \theta_i\right) = \frac{\pi_{jdi}^{(n)}[\lambda_C p^{(n)}(w|\theta_i) + (1-\lambda_C) p^{(n)}(w|\theta_{ij})]}{\sum_{\theta_u \in \Theta} \pi_{jdu}^{(n)}[\lambda_C p^{(n)}(w|\theta_u) + (1-\lambda_C)^{(n)}(w|\theta_{uj})]} \tag{4}$$

$$P\left(Z_{d,C_j,\theta_i,w} = com\right) = \frac{\lambda_C p^{(n)}(w|\theta_i)}{\ddot{e}_C p^{(n)}(w|\theta_i) + (1-\lambda_C) p^{(n)}(w|\theta_{ij})} \tag{5}$$

$$P\left(Z_{d,C_j,w} = \theta_B\right) = \frac{\lambda_B p(w|\theta_B)}{\lambda_B p(w|\theta_B) + (1-\lambda_B)[\sum_{\theta_u \in \Theta} \pi_{jdu}^{(n)}[\lambda_C p^{(n)}(w|\theta_u) + (1-\lambda_C) p^{(n)}(w|\theta_{uj})]} \tag{6}$$

$$\pi_{jdi}^{(n+1)} = \frac{\sum_{w \in d} c(w,d) P\left(Z_{d,C_j,w} = \theta_i\right)}{\sum_{\theta_u \in \Theta} \sum_{w \in d} c(w,d) P\left(Z_{d,C_j,w} = \theta_u\right)} \tag{7}$$

$$P^{(n+1)}(w|\theta_i) = \frac{\sum_{j=1}^{m} \sum_{d \in C_j} c(w,d) [1 - P\left(Z_{d,C_j,w} = \theta_B\right)] P\left(Z_{d,C_j,w} = \theta_i\right) P\left(Z_{d,C_j,\theta_q,w} = com\right)}{\sum_{w'=1}^{|V|} \sum_{j=1}^{m} \sum_{d \in C_j} c(w,d) [1 - P\left(Z_{d,C_j,w} = \theta_B\right)] P\left(Z_{d,C_j,w} = \theta_i\right) P\left(Z_{d,C_j,\theta_i,w} = com\right)} \tag{8}$$

$$P^{(n+1)}(w|\theta_{ij}) = \frac{c(w,d) \left[1 - P\left(Z_{d,C_j,w} = \theta_B\right)\right] P\left(Z_{d,C_j,w} = \theta_i\right) \left[1 - P\left(Z_{d,C_1,\theta_i,w} = com\right)\right]}{\sum_{w'=1}^{|V|} c(w,d) \left[1 - P\left(Z_{d,C_j,w} = \theta_B\right)\right] P\left(Z_{d,C_j,w} = \theta_i\right) \left[1 - P\left(Z_{d,C_j,\theta_i,w} = com\right)\right]} \tag{9}$$

3.6   Explore Across-Discipline Trendy Technology

Based on previously defined, the background theme will gather the general words and the common theme will collect words that appear permanently. And we proposed to identify inter-discipline words set, so our study will initially conduct analysis of words collected from common theme.

4   Empirical Analysis

The objective of this empirical experiment is to analyze common theme across disciplines in the abstracts of patent document. We collected patent documents from WEBPAT Taiwan, and processed by the CCMM method which calculated the probability of words appeared permanently according to the announced year of patent document and word information, furthermore, using the background model to discard too-frequent words, and the intensity analysis of the words of the common theme.

4.1   Data Collection and Extraction

In this study, four categories patent documents (IPC classification code: H04N, H04B, H04L, and G06Q) of the WEBPAT Taiwan were collected [16]. These four categories of patent documents classification are described as follows:

1. G06Q: The data processing systems or methods applies specifically to the purpose of administration, management, commerce, operation, supervision, or prediction; and the same not included in other categories.
2. H04L: Transmission of digital information, such as telegraph communications.
3. H04N: Image communication, such as TV.
4. H04B: Transmission.

We collected abstracts of these four categories patent documents collection, 1562 patent document abstracts were selected and released from 2003 to 2008, $C = \{d_1, d_2, ..., d_{1562}\}$. For analysing pervasive technology evolution, this study made the time of a year as a time interval. 2003: 378 cases; 2004: 253 cases; 2005: 324 cases; 2006: 275 cases; 2007: 265 cases; and in 2008, for the open time limit of patent information is publicly available, there are only 67 cases. Total cases of patent documents under each category: G06Q: 379 cases, H04L: 448 cases, H04N: 396 cases, and H04B: 339 cases. After collation, the whole information of patent document collections described as follows:

1. The IPC code of patent document: record the IPC classification code of patent documents.
2. Time interval (m=6): the documents were divided into 6 time intervals, i.e., $C = C_1 \cup C_2 \cup C_3 ... \cup C_6$.
3. The patent abstract: The abstract of each document was composed of words from the collection d = { $w_1, w_2, ..., w_{/V/}$ }.

4.2   Chinese Word Segmentation System Processing and Initial Value Setting

We adopted Academia Sinica's Institute Chinese word segmentation system in an SQL-based system which we developed to process batches files of Chinese patent documents. 1562 cases abstracts were fed to the Chinese word segmentation batches system and calculate the number of times (c(w, d)) the words appeared in each document. The resulting formed a two-dimensional matrix, except 1562 documents, there would collect 6158 words.

Based on our study, the sources of patent documents collecting had four categories, so we had defaulted four themes, and released six years making a total of 24 specific themes. The initial values of each cluster were set as follows:

$$
\begin{cases}
\pi^0_{jdi} = 1 \ , if \ d \in \theta_{i,j} \\
\pi^0_{jdi} = 0 \\
\pi^0_{jdi} = 0.7, if \ d \in C_j \\
\pi^0_{jdi} = 0.1
\end{cases}
\tag{10}
$$

$$
\sum_{j=1}^{6} \pi_{j,d,i} = 1
$$

$$
p^0(w|\theta_i) = \lambda_c \times \frac{\sum_{j=1}^{m} c(w,d)}{\sum_{w'=|V|} \sum_{j=1}^{m} c(w',d)}
\tag{11}
$$

$$
p^0(w|\theta_{ij}) = (1 - \lambda_c) \times \frac{c(w,d)}{\sum_{w'=|V|} c(w',d)}
\tag{12}
$$

4.3   Processing Cross-Collection Mixture Model

The CCMM used a probability distribution for words collection, which consisted of one θ_B (background model), four groups of θ_i (common theme), and 4×6 groups of θ_{ji} (specific theme), whereas it must set initial value of λ_B and λ_c by the researcher. This study set the model to gather words in θ_B with the weight as $\lambda_B = 0.95$. As most patent documents in this study included 300-500 words, with scattered words--the more frequent words were only concentrated in few words, we set a larger λ_B value. And referred to the setting λ_c of related work, we set $\lambda_c = 0.4$.

4.4   Parameters Estimation with EM Algorithm

Through the initial value setting (formula (10)-(12)) and EM algorithm (formula (3)-(9)), our study established table 1 for across technology analysis.

Table 1. π ij value of top 5 documents

Doc_id	π_{11}	π_{12}	π_{13}	π_{14}	π_{15}	π_{16}
2003_G06Q_50	0	0	0	0	0	0
2003_H04N_309	0	0.039417	0	0.97763	0.037201	0.948473
2004_H04L_510	0	0	0	0	0	0
2005_H04L_803	0	0	0	0	0	0
2006_G06Q_977	0	0	0	0	0	0
Doc_id	π_{21}	π_{22}	π_{23}	π_{24}	π_{25}	π_{26}
2003_G06Q_50	0	0	0	0	0	0
2003_H04N_309	0	0	0.003192	0	0	0
2004_H04L_510	0	1	0	0	0	0.808137
2005_H04L_803	1	1	1	0.286392	1	1
2006_G06Q_977	0	0	0	0	0	0
Doc_id	π_{31}	π_{32}	π_{33}	π_{34}	π_{35}	π_{36}
2003_G06Q_50	0	0	0	0	0	0
2003_H04N_309	1	0.960583	0.996808	0.02237	0.962799	0.051527
2004_H04L_510	0	0	0	0	0	0
2005_H04L_803	0	0	0	0	0	0
2006_G06Q_977	0	0	0	0	0	0
Doc_id	π_{41}	π_{42}	π_{43}	π_{44}	π_{45}	π_{46}
2003_G06Q_50	1	1	1	1	1	1
2003_H04N_309	0	0	0	0	0	0
2004_H04L_510	1	0	1	1	1	0.191863
2005_H04L_803	0	0	0	0.713608	0	0
2006_G06Q_977	1	1	1	1	1	1

4.5   Explore Inter-Discipline Trendy Technology

Table.2 shows $P(w|\theta_i)$ (i = 1, 2, 3, 4) of the top ten words. According to the original model common theme ($P(w|\theta_i)$) can be used to describe the clustering of the themes, thereby locating the high strength words of the common theme to help describe the theme. The keywords of common theme1(θ_1) are the words related to transmission, common theme2 (θ_2) are the words related transmission of digital information, common theme3 (θ_3) are the words for image communication, and θ_4 common theme4 are the words related to the data processing systems or methods. All of these words continue to appear and have the power of identification, so we can base on the description of each theme and filter out required keywords.

Table 2. Top ten key words of each theme

Theme1	Theme2	Theme3	Theme4
分集(Diversity)	無線裝置(Wireless devices)	位元率(Bit rate)	合會(RCAs)
符號串(Symbol string)	顯析(Significant analysis)	鏡筒(Barrel)	鑑價(Appraisal)
扇區(Sector)	解調變(Demodulation)	壓縮器(Compressor)	薪資(Salary)
多工符號(Multi-Symbol)	密鑰(Key)	鏡筒座(Seat tube)	器材(Equipment)
偏差(Deviation)	從屬(Subordinate)	標靶(Target)	工件(Workpiece)
因素(Factor)	處置器(Processor)	集成(Integrated)	代表圖(Represented Figure)
訊息碼(Message code)	資料區塊(Data block)	影像感(Image sensor)	電子會(Electrons)
遠端台(Remote station)	移動站(The mobile station)	畫像(Portrait)	產能(Capacity)
頻帶(Frequency band)	訊框組(Frame Group)	方形(Square)	前記(Pre-recorded)
附接點(Attachment points)	資料組(Data Group)	寫碼(Write code)	融資(Financing)

Since we have extracted popular words in different theme, we can cross analyze common words by intersection analysis. For example, under the condition of top 75% words of each theme, our experiments extract common words across two disciplines—transmission and transmission of digital information from 2003 to 2008, and

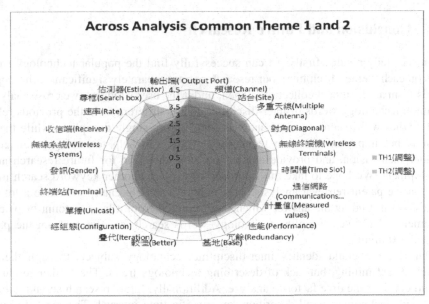

Fig. 3. Across Analysis Common Theme 1 and 2

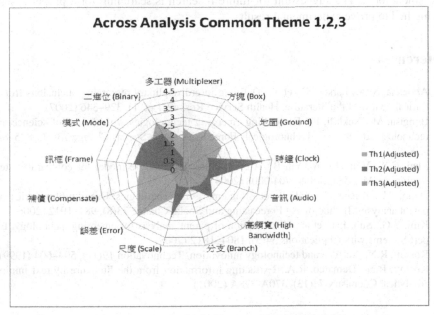

Fig. 4. Across Analysis Common Theme 1, 2 and 3

plot radar chart (Fig.3) of the result. Furthermore, under the condition of all words in each theme, our experiments extract common words across three disciplines—transmission, transmission of digital information, and image communication from 2003 to 2008, and plot radar chart (Fig.4) of the result.

5   Conclusion and Future Research

From the study results, firstly we can successfully find the popular technology subjects on each theme (discipline), our research extract separately significant technology words from all document collections. Then, under the user's request, we cross analyze common technology words of inter-discipline that is different from the previous relevant studies with patent retrieval analysis focusing on single discipline, while these obvious but important common technology words can help technology managers broader their vision, and provide another possible direction for future research and development. We hope to take this approach to explore another keyword search method on the patent retrieval analysis. This method can help enterprises save a lot of labor costs in finding the answer which they required from a large amount of patent documents. The inter-discipline technology words search is different from the past general text mining.

Our research could identify inter-discipline technology subjects through theme model of text mining, but lack of describing technology trend. The further research should consider the time factor to analyze. Additionally, future research should intend to provide technology trend describing on specific time interval. The most difficult part of text mining research is the evaluation of experimental result, because lack of assessment tools. The suggestion for future research is searching for a proper assessing method to prove experimental result.

References

1. Aboelela, S.W., Larson, E., et al.: Defining interdisciplinary research: conclusions from a critical review of the literature. Health Service Res. 42(1 pt. 1), 329–346 (2007)
2. Bengisu, M., Nekhili, R.: Forecasting emerging technologies with the aid of science and technology databases. Technological Forecasting and Social Change 73(7), 835–844 (2006)
3. Zhai, C.X., Velivelli, A., Yu, B.: A cross-collection mixture model for comparative text mining. ACM 1-58113-888-1/04/0008 (2004)
4. Daim, T.U., Rueda, G., et al.: Forecasting emerging technologies: Use of bibliometrics and patent analysis. Technological Forecasting and Social Change 73(8), 981–1012 (2006)
5. Kim, Y.G., Suh, J.H., et al.: Visualization of patent analysis for emerging technology. Expert Systems with Applications 34(3), 1804–1812 (2008)
6. Kostoff, R.N.: Science and technology innovation. Technovation 19(10), 593–604 (1999)
7. Kostoff, R.N., Demarco, R.A.: Extracting information from the literature by text mining. Analytical Chemistry 73(13), 370A-378A (2001)

8. Lee, H.J., Lee, S., et al.: Technology clustering based on evolutionary patterns: The case of information and communications technologies. Technological Forecasting and Social Change 78(6), 953–967 (2011)
9. Lee, S., Yoon, B., et al.: Business planning based on technological capabilities: Patent analysis for technology-driven road-mapping. Technological Forecasting and Social Change 76(6), 769–786 (2009)
10. Lee, S., Yoon, B., et al.: An approach to discovering new technology opportunities: Keyword-based patent map approach. Technovation 29(6-7), 481–497 (2009)
11. Mei, Q., Zhai, C.X.: A Mixture Model for Contextual Text Mining. ACM 1-59593-339-5/06/0008 (2006)
12. OuYang, K., Weng, C.S.: A new comprehensive patent analysis approach for new product design in mechanical engineering. Technological Forecasting and Social Change 78(7), 1183–1199 (2011)
13. Park, H., et al.: A Patent intelligence system for strategic technology planning. Expert Systems with Applications (2012)
14. Wu, Y., Ding, Y., Wang, X., Xu, J.: Topic Detection by Topic Model Induced Distance Using Biased Initiation. In: Kim, T.-h., Adeli, H. (eds.) AST/UCMA/ISA/ACN 2010. LNCS, vol. 6059, pp. 310–323. Springer, Heidelberg (2010)
15. Morinaga, S., Yamanishi, K.: Tracking Dynamics of Topic Trends Using a Finite Mixture Model. ACM (2004), 1-58113-888-1/04/0008
16. Shih, M.-J., et al.: Discovering competitive intelligence by mining changes in patent trends. Expert Systems with Applications 37(4), 2882–2890 (2010)
17. Son, C., Suh, Y., et al.: Development of a GTM-based patent map for identifying patent vacuums. Expert Systems with Applications 39(3), 2489–2500 (2012)
18. Tseng, Y.H., et al.: Text mining techniques for patent analysis. Information Processing & Management 43(5), 1216–1247 (2007)
19. Yoon, B., Park, Y.: A text-mining-based patent network: Analytical tool for high-technology trend. The Journal of High Technology Management Research 15(1), 37–50 (2004)
20. Zhai, C.X., Velivelli, A., Yu, B.: A cross-collection mixture model for comparative text mining. ACM 1-58113-888-1/04/0008 (2004)
21. R&D, Wikipedia,
    http://en.wikipedia.org/wiki/Research_and_development
22. The, EU Survey on R&D Investment Business Trends (2012)
23. http://wbc-inco.net/object/document/10544

A LDA Feature Grouping Method for Subspace Clustering of Text Data

Yeshou Cai[1], Xiaojun Chen[2],
Patrick Xiaogang Peng[2], and Joshua Zhexue Huang[2]

[1] Shenzhen Key Lab of High Performance Data Mining,
Shenzhen Institutes of Advanced Technology, Chinese Academy of Sciences, China
cai.yeshou@gmail.com
[2] Shenzhen University, China
{xjchen.hitsz,zhexuehuang}@gmail.com, patrickpeng@126.com

Abstract. This paper proposes a feature grouping method for cluster-ing of text data. In this new method, the vector space model is used to represent a set of documents. The LDA algorithm is applied to the text data to generate groups of features as topics. The topics are treated as group features which enable the recently published subspace clustering algorithm FG-k-means to be used to cluster high dimensional text data with two level features, the word level and the group level. In generat-ing the group level features with LDA, an entropy based word filtering method is proposed to remove the words with low probabilities in the word distribution of the corresponding topics. Experiments were con-ducted on three real-life text data sets to compare the new method with three existing clustering algorithms. The experiment results have shown that the new method improved the clustering performance in comparison with other methods.

Keywords: text clustering, subspace clustering, FG-k-means, LDA.

1 Introduction

The vector space model is commonly used in representation of text data in text mining. In this model, a document is represented as a vector in which each element is a measure of a word or a term in the set of documents. The fact that there exist thousands of unique words or terms in documents makes text data very high dimensional. The problem of sparsity always comes along with high dimensions. Real-life text documents also contain various topics and a particular document often relates to several topics, each being described as a subset of words in the document vocabulary.

Since clusters in high dimensional sparse data often occur in subspaces, many subspace clustering algorithms have been proposed to solve the high dimensional data clustering problem[1–13] . Soft subspace clustering algorithm is one cate-gory. Such algorithms assume that each feature has a feature weight to indicate its importance and subspace clusters are categorized by the subsets of features

M. Chau et al. (Eds.): PAISI 2014, LNCS 8440, pp. 78–90, 2014.
© Springer International Publishing Switzerland 2014

with larger weights. For very high dimensional data with thousands of features like text data, the current soft subspace clustering algorithms suffer that the subspaces of clusters are hardly identifiable from the feature weights because too many features are weak and sparse. To remedy this problem, FG-k-means algorithm (FGKM) [12, 13] was recently proposed to make use of two level features to generate subspace clusters, one level consisting of the individual features in data and one level composed of group features that are integration of the individual features. FG-k-means computes the weights for two level features in the clustering process to identify subspace clusters. The subspace of group features categorizes the general silhouette of the cluster which is more stable, whereas the subspace of individual features shows the detail characteristics of subspace clusters. Comparison studies have shown that FG-k-means had performance advantages over one level feature weighting in high dimensional data clustering.

In using FG-k-means, one requirement is that the two level features must be available in data. However, most real life data sets do not have such two level features. This situation greatly limits the use of the FG-k-means algorithm. To our best knowledge, there is no general approach that can be used to create two level features from one level feature data.

In this paper, we propose to use Latent Dirichlet Allocation (LDA) [14] algorithm to generate group features from text data represented in the vector space model. Latent Dirichlet Allocation method [14] is based on the assumption that a document is representing a combination of several topics and a topic is represented by a set of words or terms in the vocabulary of the document data. It has been successfully used in applications of different fields [15, 16]. The topics in the LDA model perfectly relate the feature groups of text data, which inspires the idea of applying LDA to generate group features for FG-k-means to cluster text data.

We present a LDA based feature grouping method for text data represented in the vector space model. In this method, words are aggregated into groups according to the probability distribution of words on topics which are generated by LDA. For better feature grouping, we propose a feature selection method based on words entropy to filter out the words whose probabilities are evenly distributed in several topics. After the group features are generated, the FG-k-means algorithm is applied to generate subspace clusters with two level features, the word features and the group features from high dimensional text data. We have evaluated this new method on three real life text data sets. The experiment results have shown that the new method outperformed other k-means based algorithms, including EWKM [6], W-k-means[4] and k-means.

The rest of the paper is organized as follows. Section 2 briefly reviews the FG-k-means algorithm for clustering high dimensional data. Section 3 presents the LDA based feature grouping k-means algorithm. Experiment results on real-life data are presented in Section 4 to demonstrate the advantages of the new algorithm. Conclusions and future research directions are given in Section 5.

2   FG-*k*-means for High Dimensional Data

FG-*k*-means [12, 13] is a soft subspace clustering algorithm for clustering high dimensional data, which is emerged from EWKM [6]. In this method, features are divided into feature groups according to their natural characteristics. There are two kinds of weights generated by FG-*k*-means in each cluster, weights for feature groups and weights for individual features. The feature group weights identify the subspace of the general silhouette of the cluster while the weights for individual features reveal the detail characteristics of the subspace cluster.

Let $D = \{x_1, x_2, ..., x_N\}$ be a set of N objects, and $V = \{v_1, v_2, ..., v_M\}$ represents M features of D. Let $G = \{G_1, ..., G_T\}$ be a set of T subsets of V, where $G_t \in V$, $G_s \in V$ and $G_t \cap G_s = \emptyset$ for $s \neq t$ and $1 \leq s, t \leq T$. Each subset in G is said to be an feature group of V in FG-*k*-means and represents one set of aggregated measurements in data.

Figure 1 shows the aggregation relation of feature groups and individual features. The data set has 12 features divided into three groups, $\mathbf{G} = \{G_1, G_2, G_3\}$, where $G_1 = \{v_1, v_3, v_6, v_8, v_{10}\}$, $G_2 = \{v_2, v_4, v_5, v_9\}$ and $G_3 = \{v_7, v_{11}, v_{12}\}$. For a text data D, $\mathbf{G}$ represents three subsets of 12 key words and each subset may describe a topic. For each document object x_i, the word can be measured with tf $\cdot$ idf [17]. The value of a topic in x_i is aggregated from the measures of the words related to the topic.

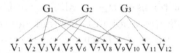

Fig. 1. Aggregation relation of feature groups and individual features

To cluster D with T feature groups into a k clusters $C = \{C_1, C_2, ..., C_k\}$, FG-*k*-means optimizes the following objective function.

$$F(U, Z, V, W) = \sum_{l=1}^{k} [\sum_{i=1}^{n} \sum_{t=1}^{T} \sum_{j \in G_t} u_{i,l} w_{l,t} v_{l,j} d(x_{i,j}, z_{l,j})$$

$$+ \lambda \sum_{t=1}^{T} w_{l,t} \log(w_{l,t}) + \eta \sum_{j=1}^{m} v_{l,j} \log(v_{l,j})] \tag{1}$$

subject to

$$\begin{cases} \sum_{l=1}^{k} u_{i,l} = 1, & u_{i,l} \in \{0,1\}, 1 \leq i \leq n \\ \sum_{l=1}^{k} w_{l,t} = 1, & 0 < w_{l,t} < 1, 1 \leq t \leq T \\ \sum_{j \in G_t} v_{l,j} = 1, & 0 < v_{l,j} < 1, 1 \leq l \leq k, 1 \leq t \leq T \end{cases} \tag{2}$$

where

- U is a $n \times k$ partition matrix where $u_{i,l} = 1$ indicates that the i-th object is allocated to the l-th cluster;
- $Z = \{Z_1, Z_2, ..., Z_k\}$ is a set of k vectors representing the centers of the k clusters;
- $V = [v_{l,j}]_{k \times m}$ is a weight matrix where $v_{l,j}$ is the weight of the j-th word on the l-th cluster;
- $W = [w_{l,t}]_{k \times T}$ is a weight matrix where $w_{l,t}$ is the weight of the t-th word group on the l-th cluster;
- $\lambda > 0$ and $\eta > 0$ are two given parameters. λ is used to adjust the distribution of W and η is used to adjust the distribution of V;
- $d(x_{i,j}, z_{l,j})$ is the distance or dissimilarity measure between object i and the center of cluster l on the j-th feature;

The above objective function can be minimized by an iterative algorithm [13] that computes the partition of objects, the cluster centers, the feature weights and the feature group weights in each iteration until the process converges. The feature weight for each feature and each cluster is computed as follows:

$$v_{l,j} = \frac{\exp\{\frac{-E_{l,j}}{\eta}\}}{\sum_{g \in G_t} \exp\{\frac{-E_{l,g}}{\eta}\}} \tag{3}$$

where

$$E_{l,j} = \sum_{i=1}^{n} \hat{u}_{i,l} \hat{w}_{l,t} d(x_{i,j}, \hat{z}_{l,j}) \tag{4}$$

and the weight for each feature group and each cluster is computed as:

$$w_{l,t} = \frac{\exp\left(\frac{-F_{l,t}}{\lambda}\right)}{\sum_{i=1}^{T} \exp\left(\frac{-F_{l,i}}{\lambda}\right)} \tag{5}$$

where

$$F_{l,t} = \sum_{i=1}^{n} \hat{u}_{i,l} \sum_{j \in G_t} \hat{v}_{l,j} d(x_{i,j}, \hat{z}_{l,j}) \tag{6}$$

When a clustering result is obtained by FG-k-means, the important topics in each cluster can be identified from $\mathbf{W}$ and the important words in each cluster can be identified from $\mathbf{V}$. This is a great benefit of FG-k-means in clustering text data if the topic groups are available. Luckily, the topic groups can be obtained with LDA algorithm.

3   LDA Feature Grouping k-means Algorithm

Latent Dirichlet allocation (LDA) is a probabilistic model to find the latent topics from a set of documents [14]. In LDA, a document is assumed to be a random mixture over Latent category-topics, where each topic is specified by a distribution over words. In this paper, we employ LDA to find the latent topics from text data.

3.1   LDA for Feature Grouping

Let $\mathbf{d}_i = \{w_1, w_2, \ldots, w_{L_i}\}$ be a sequence of L_i words in document d_i. Let $\mathbf{w}_j = \{w_j^1, w_j^2, \ldots, w_j^V\}$ for $1 \leq j \leq L_i$ in d_i be a V-dimensional vector where w_j^k $(1 \leq k \leq V) \in \{0, 1\}$, under constraint $\sum_{k=1}^{V} w_j^k = 1$. For example, the kth word in the vocabulary is represented by a V-vector $\mathbf{w}$ such that $w^k = 1$ and $w^u = 0$ for $u \neq k$. T is the number of topics to be generated from documents in corpus.

In the LDA model, each document in corpus is represented as a probability distribution over T topics. Each topic will have its own distribution over a fixed vocabulary of corpus. For example, if the topic is about computer, then the words like "hardware" and "software" will have higher probability to occur than the words like "piano" and "football". When writing a document, a person will have some topics in mind, and to explain certain topic, the person will pick one word based on the probability distribution over corpus lexicon of that topic. This hypothesis means that words occur in the same topic with high probability may express the same theme, therefore, they can be put into one group. Figure 2 is a simple example to show the relationship of words and topics. The data set has a vocabulary of 12 words. With the LDA model, we can aggregate the vocabulary into 3 topics of word groups, word group 1 about computer, word group 2 about sport, and group 3 about music.

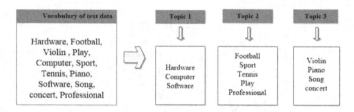

Fig. 2. Example of LDA feature grouping

Given a text data set, the LDA probabilistic model can be solved using EM algorithm[14] or Gibbs sampling [18]. The solution contains two groups of parameters, the document-level parameters Θ and the corpus-level parameters β. Θ is an $N \times T$ matrix where each row represents the probability distribution of a document d on T topics. The corpus-level parameter β represents the probability distribution of vocabulary in each topic, where $\beta_{ij} = p(v_j \mid topic = i)$

represents the probability of word v_j appearing in topic i. The elements in β satisfy $\sum_{j=1}^{M} \beta_{ij} = 1$. Parameter β can be used to summarize the theme of topics.

For each document d, we can assume there exists an $L_d \times T$ matrix ϕ_d whose element $\phi_{dji}(1 \leq j \leq L_d, 1 \leq i \leq T)$ is the probability of the j-th word in d occurring in the i-th topic with constraints $\sum_{i=1}^{T} \phi_{dji} = 1$. The T-dimensional vector ϕ_{dj} shows the probability distribution of word w_j on T topics in document d. The parameter Φ of all documents and β have the following relation.

$$\beta_{ij} \propto \sum_{d=1}^{N} \sum_{n=1}^{L_d} \phi_{dni} w_{dn}^j. \tag{7}$$

As the feature groups in the FG-k-means algorithm are disjoint but the sets of words in different topics are not disjoint. We have to make the words in different topics disjoin through word selection. Using β, we can select words for a topic as follows. Assign word $w_j \in V$ to topic G_k if $\beta_{kj} > \beta_{ij}$ for $1 \leq i, k \leq T$ and $i \neq k$. A word in the vocabulary of corpus may occur in several topics but we only put it in one topic with the largest probability β. If the probabilities of the word in two topics are equal, we can randomly put the word in one topic. However, such selection of word seems too arbitrary. In the following section, we introduce a better word selection method for selecting words in topics.

3.2 Group Feature Selection

Let n_j^t be the number of times word w_j appears in topic c_t, and f_j^t denotes the frequency of w_j in c_t. The entropy $E(w_j)$ of word w_j is given by [19]:

$$E(w_j) = - \sum_{k=1}^{T} f_j^t \times log(f_j^t) \tag{8}$$

$$f_j^t = \frac{n_j^t}{\sum_t n_j^t} \tag{9}$$

From (8), we can see that if word w_j occurs only in one topic, its entropy is 0, which means high efficiency in categorization. As the frequencies of w_j in all topics become similar, its entropy tends to be large. As such, w_j will lose its ability in categorizing topics. From this observation, we can use the entropy of a word as a measurement in word selection for topics. The parameters Φ generated by LDA are used to calculate the frequencies of words in topics. We denote Υ_d as a $T \times L_d$ matrix for document d. Each element Υ_{dij} represents the number of times that w_j in document d occurs in the i-th topic. Υ_{dij} is calculated as follows:

$$\Upsilon_{dij} = \sum_{n=1}^{L_d} \phi_{dni} w_{dn}^j \tag{10}$$

and

$$\Upsilon_{ij} = \sum_{d=1}^{N} \Upsilon_{dij} \tag{11}$$

Matrix $\Upsilon = \{\Upsilon_{ij}\}$ in (11) is the sum of the N matrices Υ_d for all documents. Let $\Psi = \{\Psi_1, \Psi_2, \dots, \Psi_M\}$. Ψ_j for $1 \le j \le M$ is the entropy of w_j calculated as follows:

$$\Psi_j = -\sum_{k=1}^{T} \left(\frac{\Upsilon_{kj}}{\sum_t \Upsilon_{tj}} \times log\left(\frac{\Upsilon_{kj}}{\sum_t \Upsilon_{tj}} \right) \right) \tag{12}$$

We sort Ψ in the ascending order and only keep the top P words with small entropies. Then, we assign each word to a unique topic.

3.3   LDA FG-k-means Algorithm

Combining the LDA topic generation model, the word selection process and the FG-k-means clustering process, we formulate the LDA based feature grouping k-means subspace clustering algorithm for high dimensional text data as given in Algorithm 1.. The algorithm consists of three major steps. Given an input text data set, the LDA process is first applied to generating LDA topic models and parameters Θ and β. Then, the document level word and topic distribution matrix ϕ_d is computed for all documents. After this, the entropies of words are computed and the words with low entropies are selected and assigned to topics. Finally, the aggregation relations of topics and selected words are obtained, and the FG-k-means clustering process is applied to the two level features to generate subspace clusters.

4   Experiment Results

In this section, we present the clustering results of the new algorithm on three real-life text data sets and compare them with the results on the same data sets by k-means, W-k-means [4] and EWKM [6]. The comparison results have shown that the new LDA FG-Kmeans algorithm outperformed other algorithms in clustering accuracy and the word selection method was specifically effective in improving the clustering performance.

4.1   Data Sets

The real life text data was the publicly available *20-Newsgroups data* set[1]. We chose the processed version of the 20news-bydate data set which was easy to read into Matlab/Octave as a sparse matrix. After filtering out the stop words,

---

[1] http://qwone.com/~jason/20Newsgroups/

Algorithm 1. LDA FG-k-means Algorithm

1: **Input:**
2: - D : Text data set;
3: - T : Number of topics;
4: - P : Percentage of words to keep;
5: - K : Number of Clusters;
6: **Method:**
7: Run LDA with T topics on D and obtain β and Φ;
8: Compute the word occurrence matrix Υ by (10) and (11);
9: Compute the vocabulary entropy vector Ψ by (12);
10: Sort Ψ in the ascending order and keep the top P percentage of words to a new set $\mathscr{V}$;
11: Group words in $\mathscr{V}$ based on β and obtain the group set $G = \{G_1, G_2, \ldots, G_T\}$
12: **for** each $d \in D$ **do**
13:     Organize the feature values of $\mathscr{V}$ into groups according to G and get a new instance dg;
14:     Adding dg in $\mathscr{D}$;
15: **end for**
16: Run FG-k-means on $\mathscr{D}$ and produce subspace clusters of Z and U;

---

we sorted the words on the inverse document frequency (IDF) and removed the words whose IDF values were not in the range of the preset upper and lower thresholds. From the master data set, we generated three text data sets with different document categories and different cluster structures in data.

Table 1 lists three data sets. N_d indicates the number of documents in each category and N_w indicates the number of words in each data set. In building each data set, we first selected the documents in specified categories from the master sparse data matrix and only kept the words/terms occurring in more than ten documents in the selected data set. We removed documents with less than nine unique words and weighed the words in each document with the "ltc" (in the SMART notation) version of $tf \cdot idf$. Finally, we normalized the weight vector of each document.

The three data sets had different cluster structures. The first data set contains two categories that were semantically different. The second data set has two categories which were semantically close. The third one has four categories and is more difficult to cluster.

As the category of each document was known, we evaluated the clustering results by calculating the correspondence between the clusters generated by a clustering algorithm and the real categories of the data set. In the experiments, precision, accuracy and F-measure were used as evaluation indices. The definition of these evaluation indices are defined in [20].

Table 1. Summary of Text Data Sets

Name	Categories	N_d	N_w
D_1	alt.atheism	472	1066
	comp.graphics	533	
D_2	talk.politics.guns	526	1465
	talk.politics.mideast	518	
D_3	misc.forsale	442	1923
	re.sport.hockey	540	
	sci.crypt	572	
	soc.religion.christian	565	

4.2   Parameter Settings

In this experiment, we investigated the proper settings of parameter T, the number of topics, and parameter P, the percentage of the words to keep. We built multiple LDA models from the three data sets with different numbers of topics $T \in \{5, 10..., 30\}$ and different $P \in \{40\%, 50\%, ...80\%, 90\%\}$. We used FG-$k$-means to cluster the data sets with different combinations of topics and words and computed the accuracy, F-measure and precision of each clustering result. Figure 3 shows the clustering performance of different settings on the three data sets measured in accuracy, F-measure and precision. The darker region in the contour map indicates better clustering results. The horizontal axis is the number of topics T and the vertical axis is P. From these figures, we can observe that the best results of data set D1 appeared with setting of $2 * k < T < 3 * k$ and $50\% \leq P \leq 80\%$ where k is the number of clusters. The best results of data set D2 occurred in $2 * k < T < 3 * k$ and $P \leq 50\%$ and the best results of data set D3 occurred in $2 * k < T < 3 * k$ and $P = 50\%$. From these observations, we can set the default value P around 50% and the default value T as two or three times of the number of clusters, i.e., between $2k$ and $3k$.

4.3   Comparisons of LDA FG-k-Means and Other Clustering Algorithms

We compared FG-k-means with the LDA topic word grouping scheme and other three clustering algorithms, i.e., k-means, W-k-means and EWKM. In topic and word grouping, we simply assigned each word in the vocabulary to the topic in which the word has the highest probability in matrix β computed with LDA. The numbers of topics for the three data sets were set as 5, 5 and 10. Other three algorithms ran on the original text data in the vector space model.

Table 2 summarizes the clustering results of the three data sets by the four clustering algorithms. The results were evaluated in three measures. For each data set, 100 clustering results were produced by each algorithm with different initializations. Each value in the right column of Table 2 is the average of 100 results by the FGKM algorithm. The value in the brackets is the standard deviation of the

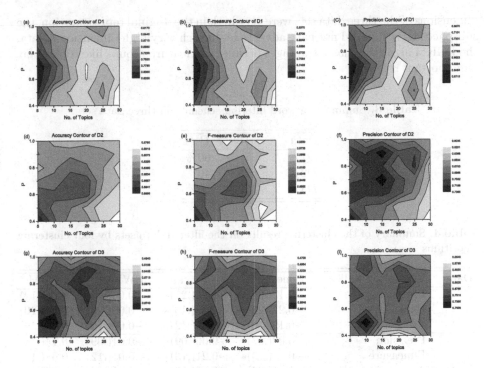

Fig. 3. (a)-(c) The contour maps of accuracy, Fmeasure and precision on D1. (d)-(f) The contour maps of accuracy Fmeasure and precision on D2. (g)-(i) The contour maps of accuracy Fmeasure and precision on D3.

Table 2. Summary of clustering results on three data sets in Table 1 by FG-k-means and other three clustering algorithms

Data	Evaluation Indices	k-means	W-k-means	EWKM	FGKM
D1	Precision	−0.05(.20)	−0.28(.01)∗	−0.10(.18)∗	0.80(.10)
	F-measure	−0.11(.24)∗	−0.20(.01)∗	−0.20(.17)∗	0.71(.15)
	Accuracy	−0.06(.17)	−0.23(.01)∗	−0.13(.11)∗	0.74(.12)
D2	Precision	−0.06(.10)∗	−0.05(.09)∗	−0.03(.11)	0.69(.09)
	F-measure	−0.08(.16)∗	−0.15(.11)∗	−0.13(.14)∗	0.60(.10)
	Accuracy	−0.03(.11)	−0.07(.07)∗	−0.06(.09)∗	0.64(.08)
D3	Precision	−0.04(.17)	−0.08(.17)∗	−0.13(.14)∗	0.70(.12)
	F-measure	−0.04(.21)	−0.15(.20)∗	−0.33(.13)∗	0.60(.11)
	Accuracy	−0.01(.16)	−0.08(.16)∗	−0.24(.11)∗	0.62(.09)

100 results. The values in other columns are the difference between the FGKM result and the result of the corresponding algorithm on the same data. "∗" indicates that the difference is significant. From these results, we can see that FGKM significantly outperformed all other clustering algorithms. Compared with W-k-means and EWKM, the k-means algorithm performed better. This is because the

dimensions of the three data sets were very high in the original representation, the subspace algorithms did not have advantage. In such very high dimensional data, the FGKM algorithm has advantages because the feature groups like topics play an important role.

Table 3. Summary of word filtering schemes on three data sets

Data	T	P
$D1_f$	5	60%
$D2_f$	5	40%
$D3_f$	10	50%

Table 4. Summary of the clustering results on the filtered data sets by four clustering algorithms

Data	Evaluation Indices	k-means	W-k-means	EWKM	FGKM
$D1_f$	Precision	−0.13(.20)*	−0.20(.21)*	−0.04(.15)	0.86(.12)
	F-measure	−0.27(.22)*	−0.35(.16)*	−0.09(.25)	0.81(.20)
	Accuracy	−0.19(.16)*	−0.25(.12)*	−0.06(.19)	0.83(.16)
$D2_f$	Precision	−0.02(.07)	−0.00(.09)	−0.01(.09)	0.74(.11)
	F-measure	−0.18(.13)*	−0.21(.13)*	−0.01(.12)	0.68(.14)
	Accuracy	−0.11(.09)*	−0.13(.09)*	−0.01(.10)	0.70(.12)
$D3_f$	Precision	−0.11(.16)*	−0.16(.14)*	−0.16(.13)*	0.75(.11)
	F-measure	−0.10(.15)*	−0.23(.16)*	−0.19(.16)*	0.68(.13)
	Accuracy	−0.07(.11)*	−0.18(.13)*	−0.13(.14)*	0.70(.11)

To investigate the effectiveness of the group feature selection described in Section 3.2, we conducted the word filtering process on the original representation of the three data sets with different P to produce three reduced data sets as shown in Table 3. We used the four clustering algorithms to cluster the filtered data sets. The results are shown in Table 4. Compared with the results in Table 2, we can see that the clustering performances of FGKM further improved significantly. These results demonstrated that the group feature selection method was indeed effective because noise words were removed from the data. In the filtered data with reduced dimensions, the subspace clustering algorithm EWKM performed much better than k-means and W-k-means which cluster data on the entire space. Although W-k-means can automatically reduce the effects of noise features, it did not work well on the filtered data because noise words were filtered out already. W-k-means cannot identify subspace clusters but EWKM can. Therefore, EWKM performed much better on the filtered data sets than W-k-means. However, since the dimensions in the filtered data sets were still high, the FGKM algorithm performed better than EWKM.

5   Conclusions

In this paper, we have proposed the LDA FG-k-means algorithm for clustering high dimensional sparse text data. In the new algorithm, we first applied LDA algorithm to generate topic models. From the topic models, we created the group level features by assigning each word in the vocabulary to one topic. Then, we used a word filtering process to remove noise and insignificant words according to the entropy of the word computed from the topic models. Finally, we used the feature grouping weighting subspace clustering algorithm FG-k-means to cluster the filtered data. Experiment results have shown that the new algorithm improved the clustering performance of high dimensional sparse text data significantly. We have also demonstrated that the word filtering method based on LDA models improved clustering results of subspace clustering. Our future work is to explore the use of LDA models in ensemble clustering of high dimensional sparse text data.

Acknowledgement. This research is supported by NSFC under Grant No. 61305059 and No.JC201005270342A.

References

1. Makarenkov, V., Legendre, P.: Optimal variable weighting for ultrametric and additive trees and k-means partitioning: Methods and software. Journal of Classification 18(2), 245–271 (2001)
2. Modha, D., Spangler, W.: Feature weighting in k-means clustering. Machine Learning 52(3), 217–237 (2003)
3. Friedman, J., Meulman, J.: Clustering Objects on Subsets of Attributes. Journal of the Royal Statistical Society Series B (Statistical Methodology) 66(4), 815–849 (2004)
4. Huang, Z., Ng, M., Rong, H., Li, Z.: Automated variable weighting in k-means type clustering. IEEE Transactions on Pattern Analysis and Machine Intelligence 27(5), 657–668 (2005)
5. Domeniconi, C., Gunopulos, D., Ma, S., Yan, B., Alrazgan, M., Papadopoulos, D.: Locally adaptive metrics for clustering high dimensional data. Data Mining and Knowledge Discovery 14(1), 63–97 (2007)
6. Jing, L., Ng, M., Huang, Z.: An entropy weighting k-means algorithm for subspace clustering of high-dimensional sparse data. IEEE Transactions on Knowledge and Data Engineering 19(8), 1026–1041 (2007)
7. Hoff, P.: Model-based subspace clustering. Bayesian Analysis 1(2), 321–344 (2006)
8. Bouveyron, C., Girard, S., Schmid, C.: High Dimensional Data Clustering. Computational Statistics & Data Analysis 52(1), 502–519 (2007)
9. Tsai, C.Y., Chiu, C.C.: Developing a feature weight self-adjustment mechanism for a k-means clustering algorithm. Computational Statistics & Data Analysis 52(10), 4658–4672 (2008)
10. Deng, Z., Choi, K.S., Chung, F.L., Wang, S.: Enhanced soft subspace clustering integrating within-cluster and between-cluster information. Pattern Recognition 43(3), 767–781 (2010)

11. Cheng, H., Hua, K.A., Vu, K.: Constrained Locally Weighted Clustering. Proc. VLDB Endow. 1, 90–101 (2008)
12. Chen, X., Xu, X., Ye, Y., Huang, J.Z.: Tw-k-means: Automated two-level variable weighting clustering algorithm for multi-view data. IEEE Transactions on Knowledge and Data Engineering 25(4), 932–944 (2013)
13. Chen, X., Ye, Y., Xu, X., Huang, J.Z.: A feature group weighting method for subspace clustering of high-dimensional data. Pattern Recognition 45(1), 434–446 (2012)
14. Blei, D.M., Ng, A.Y., Jordan, M.I.: Latent dirichlet allocation. In: Proceedings of the Twenty-Second Annual International SIGIR Conference, vol. 3(4-5), pp. 993–1022 (2003)
15. Griffiths, T.L., Steyvers, M.: Finding sientific topics. Proc. Natl. Acad. Sci. U.S. A 101(suppl. 1), 5228–5235 (2004)
16. Bhattacharya, I., Getoor, L.: A latent dirichlet model for unsupervised entity resolution. In: SIAM Conference on Data Mining (SDM), pp. 47–58 (April 2006)
17. Salton, G., Buckley, C.: Term weighting approaches in automatic text retrieval. Information Processing and Management 24(5), 513–523 (1988)
18. Porteous, I., Newman, D., Ihler, A., Asuncion, A., Smyth, P., Welling, M.: Fast collapsed gibbs sampling for latent dirichlet allocation. In: KDD 2008 Proceedings of the 14th ACM SIGKDD International Conference on Knowledge Discovery and Data Mining (2008), pp. 569–577 (2008)
19. Shannon, C.E.: A mathematical theory of communication. Bell System Technical Journal 27, 379–423, 623–656 (1948)
20. Sokolova, M., Lapalme, G.: A systematic analysis of performance measures for classification tasks. Information Processing and Management 45, 427–437 (2009)

Author Index